Intentionality and Action

Edited by
Jesús Padilla Gálvez and Margit Gaffal

DE GRUYTER

ISBN 978-3-11-055909-5
e-ISBN (PDF) 978-3-11-056028-2
e-ISBN (EPUB) 978-3-11-055910-1

Library of Congress Cataloging-in-Publication Data
A CIP catalog record for this book has been applied for at the Library of Congress.

Bibliographic information published by the Deutsche Nationalbibliothek
The Deutsche Nationalbibliothek lists this publication in the Deutsche Nationalbibliografie; detailed bibliographic data are available on the Internet at http://dnb.dnb.de.

Printing and binding: CPI books GmbH, Leck
♾ Printed on acid-free paper
Printed in Germany

www.degruyter.com

FSC www.fsc.org MIX Papier aus verantwortungsvollen Quellen FSC® C083411

Table of Contents

Jesús PADILLA GÁLVEZ and Margit GAFFAL

Intentionality and Action

An Introduction

Since antiquity philosophers have been concerned with certain problems that have lost none of their topicality in the course of history. One of these problems still to be solved is the thought-figure of Aristotle's practical syllogism. The practical syllogism consists of a major and a minor premise and a conclusion in the form of an action.[1] There has been an ongoing discussion about the structure of these premises. It was not until the publication of Franz Brentano's writings that the logical configuration of the premises was reconstructed. In order to define the practical argument we need to introduce the concept of intention. Although the notion of intentionality goes back to the Middle Ages, its structure was still obscure and lacked a clear definition at that period.[2] Brentano described 'intentionality' as consisting of mental processes and acts of consciousness and it reads as follows:

> Every mental phenomenon is characterized by what the Scholastics of the Middle Ages called the intentional (or mental) inexistence of an object, and what we might call, though not wholly unambiguously, reference to a content, direction towards an object (which is not to be understood here as meaning a thing), or immanent objectivity. Every mental phenomenon includes something as object within itself, although they do not all do so in the same way. In presentation something is presented, in judgement something is affirmed or denied, in love loved, in hate hated, in desire desired and so on.
> This intentional in-existence is characteristic exclusively of mental phenomena. No physical phenomenon exhibits anything like it. We could, therefore, define mental phenomena by saying that they are those phenomena which contain an object intentionally within themselves.[3]

The question of the content of an intentional proposition marks the starting point for our considerations. In other words: what does intention refer to and what is its substance? Brentano explained that mental states have certain char-

1 Aristoteles, *Nicomachean Ethics*, 1141b 14–20. Aristoteles, *De Motu Animalium*, 701a 12–14.

2 Ibn Sina, *Kitab al-Isharat wa-'l-tanbihat* (*Remarks and Admonitions*). Ed. S. Dunya, Cairo, 1960, Vol. II, 343–346. Domenicus Gundissalinus, *De Processione Mundi*. Eds. M. J. Soto Bruna, C. Alonso del Real. Pamplona, Eunsa, 1999. Thomae Aquinas, *Summa Contra Gentiles*. Edition Leonina manualis. Torino, Marietti, 1948.

3 Franz Brentano, *Psychology from an Empirical Standpoint*. Ed. Linda L. McAlister. London, Routledge, 1995, 88f.

https://doi.org/10.1515/9783110560282-007

acteristics and are directed to something which has a representational or semantic content. Mental states that involve intention are generally referred to as intentional states. But the question arises as to how these intentional states are naturalized? Over the last century philosophers proposed several answers to these questions from divergent perspectives: in phenomenology, for instance, sensations and experience are regarded as intentional qualities.[4] The same as in phenomenology and partly in the analytical tradition, Brentano defined intentionality by certain characteristic features and saw it located in the mental domain.

The logical thought of practical syllogism had to be redefined.[5] Suppose an agent A has the intention to achieve a particular result. A studies hard in order to achieve good grades and having done so, A receives a reward. The first premise describes what the agent must do to achieve the goal. We are interested in the reasons for A's success and ask for A's propositional attitude. A may have reached his goal using such techniques as, for instance, he "has learned to cheat" or he has memorized very single detail of a subject matter, etc. The conclusion would then be a clear definition of the possible result and the content of intention. If A achieves good grades he gets a reward. The basic argument has the following form:

> Major premise: A is studying hard to achieve good grades; (*intention*)
> Minor premise: A believes that by studying hard he can reach the goal; (*propositional attitude*)
> Conclusion: If A has good grades, he gets a reward. (*action*)

It could, however, as well happen that despite of A's effort he fails the test and is denied the reward. Or reversely, A cheated and without having studied the subject matter he gets good grades and receives the reward, etc. These possibilities give rise to two questions that must be clarified: first, what is the logical status of the conclusion? And second, given the logical validity of the conclusion, what scheme of comprehending behavior is presented for an agent's intention and cognitive attitude? Before answering these questions we need to take an intermediate step in which we operationalize Brentano's proposal.

---

4 Edmund Husserl, *Logische Untersuchungen. Zweiter Teil: Untersuchungen zur Phänomenologie und Theorie der Erkenntnis*. Niemeyer, Halle 1901, cf.: V. Über intentionale Erlebnisse und ihre "Inhalte", 322 ff.

5 Aristoteles, *Nicomachean Ethics*, 1141b 14–20.

It was Gottlob Frege who put a limit to the mental domain, as he considered the content of a proposition (*thoughts*) as most relevant.[6] For him the word 'thought' denoted a propositional content rather than a psychological process, as was presupposed by Brentano. Frege's thesis does not refer to mental processes but states that a comprehensive description of a propositional content can only be achieved by analyzing the language. The underlying conviction is that propositional content cannot be determined by ideas or other components that form part of a stream of consciousness. In contrast to the view that thoughts belong to a person's individual stream of consciousness and can never be clearly known by others, it is held that the same propositional content can equally be known by different persons. According to Brentano, intentionality is something individual and can be attributed to a person given the person's singular reasoning and train of thoughts. On the contrary, we believe that intentionality has a propositional content because it was learned publicly and is mediated by language.

In his further studies Frege placed the expression of thoughts at the focus of interest. He aimed to reveal the logical relations among thoughts. Thus he examined the sentence structure and all its possible meanings and contrasted the manifold ways of expressing one and the same proposition (thought). In doing so, he achieved a change of paradigm because he placed the philosophical logic of thoughts at the center of attention. This enabled him to liberate the proposition from a purely inner space of consciousness which permitted him to objectify it. Wittgenstein took up Frege's perspective and made it his task to overcome the methodological solipsism. Thinking was no longer viewed as something purely inward, isolated and individualized. Our actions were not developed in a claustrophobic inner space to find their expression in form of intentions. What we call mental processes are interactions of language and environment that are intertwined in manifold ways. Three conclusions can be drawn from this: first, as the content of our actions take place externally, the content of our intentions is also determined by the external world. Second, as our linguistic use is referential, our intentions are also constituted by our interactions with the environment. And third, if our language games are closely related to our forms of life, then we have acquired relevant parts of our thinking externally, including our views and intentions. The assumption that mental processes are independent of language and are processed in a retired inwardness is implausible and must be

---

6 G. Frege, Der Gedanke. Eine logische Untersuchung, *Beiträge zur Philosophie des deutschen Idealismus*. Vol. I, 1918–1919, 58–77.

questioned. Thus, the position that Brentano assigned to phenomenology and to intentional acts has to be re-considered and possibly put within a new frame.

Wittgenstein's novel approach is based on the view that propositions do not require any specific mentality. If a person sees that it rains and thinks "It's raining", then the truth condition of the thought is just nothing mental. When I see gray clouds and feel a fresh wind blowing and I wish it would rain, this desire is not a mental act. And the desire cannot be explained by describing a mental state. This position is opposed to Brentano's thesis that the content of thoughts is of vital importance and determinant for the linguistic content. Therefore Brentano characterized intentionality directly without the indispensable detour of its linguistic manifestation.

The problem of intention is usually investigated from three perspectives: first, intention is seen as an action directed towards the future; second, intention is ascribed to somebody and third, intentional actions are analyzed. Each of these facets points to a different problem and involves a particular mode of expression. When trying to give an overall definition of intention we realize the difficulty of linking it to a reliable criterion that is true for every intentional act. For instance, the claim that intention is always future-oriented could be opposed by objecting that actions may also be called intentional without having anything to do with future.[7]

The main task of philosophy is to discover and describe the different forms by which intention is expressed. This undertaking has not only been important for the philosophy of mind but also for ethics and its implications for epistemology and practical reasoning. In recent history philosophers have made several proposals regarding intention, one of which was suggested by Donald Davidson. He suggested a reductive theory of intention as "synchronization" according to which intentional propositions do not refer to an event or an agent's state of mind but to a re-description of what is being done.[8] The relation between an agent's action and the favorable attitude towards his action is what is considered intentional. Thereby Davidson combined intention with action. However, this proposal cannot incorporate an intentional action directed towards the future.[9] There are apparent cases of what may be called 'pure intention' in which no measures of any kind are taken. Suppose I intend to write a

7 G. E. M., Anscombe, 1963, *Intention*, 2nd edition. Oxford, Blackwell, 1.

8 D. Davidson, 1963, 'Actions, Reasons, and Causes', reprinted in *Essays on Actions and Events*. Oxford, Oxford University Press, 1980, 3–20. Cf.: 5–8.

9 D. Davidson, 1978, 'Intending,' reprinted in *Essays on Actions and Events*. Oxford, Oxford University Press, 1980, 83–102.

book review although I have not yet opened the book.[10] At first glance, this pure intention cannot be reduced to an intentional action. But if we recognize a pure intention, there is no reason why there could not be another equally relevant intention during the time of the intended action. If what I am doing intentionally requires time, as almost everything does, there will be early phases at which I am aware that I have to write the review and therefore take action, though I have not yet begun.

The state of a prospective intention cannot be reduced to a single action, which is why we need other intentional forms, expressed as an endeavour or an attempt as part of a mental state. This perspective is clearly rejected by philosophers who are inspired by Anscombe's view of "intention". If someone is asked "Why are you doing A?", Anscombe considered a distinction between the following two answers unnecessary: "I am doing A" and "I am going to do A".[11] She claimed that once an intention to a future action has been understood no further explanation is needed. As such, Anscombe sought to solve the problem of unity without recognizing intention as a mental state.

If no distinction is made between the two possibilities mentioned above then both enunciations refer to an open future. A statement such as "He is doing A" does not imply that the agent will succeed in doing A, nor that he has reached an aim or will get very far with it. We consider the fact that S intends to do A as synonymous with the fact that S is doing A intentionally, even though S has not yet commenced A. In other words, if I intend to visit Toledo next Thursday, I am on my way to do so. Here it is surprising that sometimes we use the progressive form for a current progress in advance although the action itself has not yet begun.

This future reference allows us to examine the different language games involved. Thus trying to do A should not be considered a mental state because the whole process of trying is in fact not static. Instead, it is rather an imperfect form of progress towards the intentional completion of an act, whereby progress can be so rudimentary, ineffective or interrupted that it would be strange to state "He is doing A." Both statements, to 'have a prospective intention' and to "do something intentionally" may be perfectly consistent as long as an action is seen in progress.

Along with this view and the use of an anticipated future there are two main arguments for the theory of intention linked to intentional action. The first argument explains why intention is always understood as an intention to do

---

10 G. H. von Wright, 1972: On So-Called Practical Inferences, *Acta Sociologica*, 15, 39–53.

11 G. E. M., Anscombe, 1963, *Intention*, 2nd edition. Oxford, Blackwell, 39–40.

something. Although intention is sometimes alleged as propositional attitude in the form “I have the intention of doing so”, this statement can always be expressed through “I intend to ...” which is an assertion that I intend to achieve p. On the contrary, it is simply nonsense to rephrase such common expressions as “I intend to go home” in propositional terms of “I have the intention of walking home”. If intention is considered to be a step on the way to intentional action, it is clear that its object is not a mere proposition but something that one can achieve.

Second, to assume that intention gives rise to intentional action may elucidate what is generally termed “rationalization”. Thus an action can be explained as involving intention if we, for instance, affirm “He is doing A, because of his intention to do B”. In such affirmation the speaker justifies an action by relating it to the agent’s past or present intention. The use of the conjunction ‘because’ leads us to think that we have developed a rational process in which facts are guided by mental processes involving an implication. The function of the subordinating conjunction ‘because’ as a rational criterion on which the link between intentionality and action is based, may help clarify the target-oriented intention. Hence if S is doing A then this action can be interpreted as an intentional-in-progress action that explains A’s action.

We have seen that in the practical syllogism there is a relation between intention and propositional attitude. We have also shown that intention is interconnected with belief. If someone intends to do A, one believes that one is going to do A. However, Paul Grice made us aware of the inconsistencies that may go along with intention, as for instance, in the following case: “I intend to do A, but I may not be able to do it.”[12] This can easily be explained if intention is a kind of belief. The same is true for a future-oriented intention such as in “I am going to do A”, which is nothing else than an affirmative expression of a belief. The same applies to an intention that is linked to action. If a speaker asks, “What are you doing?” and receives the response “I am doing A”, then the link between intention and belief seems evident even though there are other possible alternatives to explain this dialogue.[13]

Two arguments were brought up against this conception. The first regards apparent cases of intention without belief. According to Davidson, “a person may be intentionally making ten copies in writing but at the same time he might

12 H.P. Grice, 1971: ‘Intention and Uncertainty’, *Proceedings of the British Academy*, 57, 263–279. See: 264–266.

13 D. Davidson, 1978, ‘Intending’, in: *Essays on Actions and Events*. Oxford, Oxford University Press, 1980, 83–102. See: 91 and 100.

not know what this is for. All he knows is that he is intending to do it.[14] There are similar cases of this kind, for instance, if I intend to send my paper to a scientific journal though I am aware that I am sometimes forgetful, and therefore might not send it. Such examples can be explained in various ways. One explanation is that when I think I am going to do something which in fact remains undone, I do not have the actual intention of doing this action but just intend to try it. The question that we should clarify is whether the whole process is only a metaphor without resulting in any action, though I lie to myself and say that I have tried it. Then the future-oriented enunciation lacks any content whatsoever.

A more radical response points to a simplifying assumption, often made in epistemology and in philosophy of action, that belief is of binary nature and does not occur gradually. It may be harmless to say that intention implies belief but we must not forget that a person's manifestation of intention may inspire confidence or a process of trust in the recipient. We may assume that an agent's expression of intention corresponds to an act of volition and implies that an action will happen in the near future. Both aspects are likely to create problems if an agent expressing an intention does not perform what he had intended but the only purpose was to inspire a state of trust in the other or to induce the other to make specific assumptions. Such strategies fail to create the necessary cognitive conditions on which intention and volition are based.

The second objection is of epistemic nature. If forming an intention is, among other things, come to believe that one is either doing or going to do A, then the question arises: what gives us the right to express such a belief? To enunciate an intention is different from predicting the future based on what is considered to be or should be the case. That is why Brentano pointed to a kind of knowledge whose content is not yet fulfilled. His proposal implies that – strictly speaking – we should not only exclude the content of intentional expressions but also any knowledge related by inference. As a consequence, if we reduce practical to theoretical reasoning and hold that intention merely implies belief, then we must doubt whether intentions are formed on the basis of previous evidence. If we postulate that beliefs lack such support then we have to regard intention as a case of fantasy situated outside any epistemic structure. Thus the enunciation "S intends to do A" could be rephrased by the enunciation "S thinks that he is going to do A". The second version lacks sufficient prior

14 D. Davidson, 1971: 'Agency,' reprinted in *Essays on Actions and Events*. Oxford, Oxford University Press, 1980, 43–61. See: 50. Cf.: D. Davidson, 1978, 'Intending,' reprinted in *Essays on Actions and Events*. Oxford, Oxford University Press, 1980, 83–102.

evidence. On the contrary, we know what we are doing, or what we are going to do by inferring the conditions of our will together with premises about our own abilities. Yet, the will does not imply any belief and therefore does not necessarily include any intention.

The articles presented in this book deal with some of the issues and problems described above from three different points of view. On the one hand the problem of intentionality is analyzed from a historic perspective and relevant approaches to the problem of intention are presented. Then the most important arguments of each approach are developed and critically discussed. The correlation between historical context, conceptual content and rational argument makes it possible to clarify a subject as complex as the relation between "intentionality and action".

The present volume deals with the interrelation between action and intentionality and comprises contributions from the following scholars. Kevin Mulligan investigates the relation between intentionality, motives and reasons. Peter Simons outlines the philosophical significance of intentionality by investigating the relation between agency and conscious intentionality. Wolfgang Künne analyses the role of the concept of intentionality in Husserl's logical investigations. Christian Bermes focuses on the meaning of commitment and compares its significance with a theory of intentionality. Jesús Padilla Gálvez discusses the possible paradoxes that intentional acts may produce if they are not clearly determined. Michel Le Du approaches the intentionality of actions by its possible dysfunctions making reference to the problem of egocentricity and the will. António Marques explores the problem of intentionality in Wittgenstein's philosophy of psychology. Margit Gaffal proposes a distinction between intention and '*Absicht*' and looks into the language games that are commonly used to express intentional acts. Severin Schroeder examines the difference between motives and causes from the first-person authority. Amir Horowitz analyses intentional relations within recent analytic discussions. Bernhard Obsieger explores the morality of actions in the scope of Husserl's ethics.

Kevin Mulligan

Intentionality, Reasons and Motives

1 Intentionality vs Reasons & Motives

Philosophies of intentionality, of the “directedness” of mental acts and states abound. The distinctions between non-propositional and propositional intentionality, between relational and non-relational intentionality, between objects and propositions or states of affairs, content and object, conceptual and non-conceptual content, sensations and content, are the focus of a variety of well-known disagreements.

The same is true of the philosophies of reasons, reasonableness and rationality, and motives. There, too, a number of distinctions are the focus of a variety of disagreements – the distinctions between motives and reasons, between defeasible and non-defeasible justification, between affective or conative motives and intellectual motives, between causality and intelligibility, between affective, practical and intellectual rationality, or reasonableness.

What is the relation between intentionality, reasons and motives? Answers to this question are, of course, determined by views about the nature of intentionality and reasons. In what follows, I outline a view of some possible relations between intentionality, motives and reasons in which the distinction between correct and incorrect mental states and of the idea of correctness-makers are prominent.

2 Reasons

Philosophers have always had a lot to say about *Reason*. They have not always had as much to say about the nature and variety of *reasons*. Recognition of the variety of reasons seems to be peculiar to twentieth century philosophy. There are reasons to *F* and reasons not to *F*. What may we substitute for “*F*”? Two popular candidates are reasons to *act* and reasons to *believe* or *judge*. But although there seem to be no good reasons to deny that there are reasons to desire, will, admire, despise, regret and resent, interest in the philosophy of reasons to pro-emote and to contra-emote seems to be very much a twentieth century phenomenon. Reason, we may think, is either theoretical, practical, conative or affective, taking for granted a particular taxonomy of mental acts and states and

https://doi.org/10.1515/9783110560282-015

actions. Is there, then, a type of mental act or state, *G*, which is such that there is no reason to *G* and no reason not to G? One good candidate for this role is knowledge. Even if knowledge is a type of belief that *p* and there are reasons to believe that *p*, the belief that *p* which knowledge is, on this popular view, correct or true belief which is justified and satisfies certain other conditions, is not itself something for which there is a reason. And if knowledge is not any type of belief it seems equally implausible to say that there could be a reason to know.

For the moment we may understand the relation of being a reason to *F* as a relation of speaking in favour of *F*-ing and the relation of being a reason not to *F* as a relation of speaking against *F*-ing. There are two distinct ways of formulating this relation. We may say that Sam has a reason at *t* in a context, *c*, to admire Maria or to propose to her or to judge that she is admirable. Call this the relativized formulation of the reason relation. But there is also a non-relativised way of formulating the reason relation which we employ when we say that someone's courage is a reason to admire her or that the injustice of some situation is a reason to be indignant about it. On one view, the reason relation should always be relativized or specified in the way mentioned and then the non-relativised formulations mentioned turn out to be elliptic formulations. But even the view that there are some reason relations which are not elliptic in this way should allow for the need for relativized reason relations. The tides in the affairs of men are such that at some time, in a particular context something is a reason for a particular person to act or feel in a certain way. There is what Goethe called a *Forderung der Stunde*. Another even more important reason for formulating the reason relation in a relativized way is, of course, that in a particular context and at a particular time there are often numerous reasons for and against *F*-ing and since these are interrelated in numerous ways a particular reason to F will only be what it is, will be individuated by, these other considerations.

If to be a reason is to be a term of one of the two relations mentioned, what sort of item can play this role? When we say that the danger a particular dog represents is a reason to be afraid of it or to kick it, it is the fact that the dog is dangerous, which is the reason which speaks in favour of being afraid of it or kicking it. This suggests that facts are the terms of the reason relation. But there are two very different ways of understanding what it is to be a fact. On one view, a fact is a true proposition, where a proposition is something like a Fregean thought. On another view, a fact is an obtaining state of affairs. According to the former view, a fact is not the sort of thing which can contain something like a dog. According to the latter view, a fact may contain items such as dogs and mountains.

In what follows, I shall not attempt to decide which view of facts is required by the claim that one of the terms of the reason relation is a fact. I shall however say something about the nature of the reason relation.

It is an internal relation. A relation is internal if, given its terms, they must stand in the relation they stand in. So runs the traditional account, which I shall presuppose here. The traditional examples of internal relations are relations holding between abstract or non-temporal items, such as colours and numbers. Thus the relations between the number 3 and the number 2, in that order, of being greater than or being the immediate successor of are internal relations. The two numbers must stand in those relations. Similarly, the relation of lying between which relates orange, red, and yellow, in that order, is an internal relation. The three colours must stand in the relation.

If the reason relation is an internal relation, it is not an internal relation of this kind. For not all internal relations are relations between non-temporal items. Acting, emoting, believing are not ideal entities. Nor are the times, contexts and subjects mentioned in relativized formulations of the reason relation. Facts, on some accounts, are not in time. But if facts are obtaining states of affairs, then they are the sort of thing which may contain temporal items such as mountains and dogs.

The reason relation satisfies the definition of internal relations: the fact that a situation is unjust and indignation about the situation cannot fail to stand in the reason relation, the fact cannot fail to be a reason to be indignant about it. But its *relata* are not of the same kind as the *relata* of the least controversial examples of internal relations, numbers and colours. This might be thought to provide an objection to the claim that the reason relation is an internal relation.

In fact, many types of internal relation relate terms all or some of which are temporal items. One example is numerical difference between temporal items. Another is provided by the relations of numerical difference and qualitative dissimilarity or similarity between tropes or particularised features.

The reason relation, I suggest, is an internal relation which, like the examples just mentioned, relates temporal items.

Reasons come in two very different kinds. They are either defeasible or non-defeasible. More exactly, if something is a reason to *F*, it is either a defeasible or non-defeasible reason to *F*. The status enjoyed by some fact when it is a defeasible and undefeated reason to *F* may be lost. If numerous normally reliable experts assure me that *p*, that is a defeasible reason for me to believe that *p*. But that reason to believe that *p* is defeated by the fact that the world authority on the matter, whose authority and reliability is accepted by all the other experts, assures me that not-*p*. Similarly, if a situation displays many of the features of an unjust situation, that is a reason to be indignant about the situation. But that reason is defeated by the fact that one necessary feature of unjust situations is not present. If a defeasible reason is a reason which may be defeated, a non-defeasible reason is one which cannot be defeated. Perhaps the unjustness of a situation is a non-defeasible reason to be indignant about it, the shamefulness of a deed a non-defeasible reason to be ashamed of the deed, and the fact that p a non-defeasible reason to believe that p.

But one may wonder whether there *are* any non-defeasible reasons. A type of scenario devised by Rabinowicz and Ronnow-Rasmussen suggests that there are no non-defeasible reasons. Suppose that the fact that some situation is unjust is a reason for you to be indignant about the situation. Suppose further that you are offered a large amount of money provided you are not indignant about the situation. Does the offer not defeat the reason to be indignant about the situation? One very plausible answer to this question, which I think I first heard from John Skorupski, goes as follows. The offer constitutes a very good reason to *act* in a certain way, to do what is necessary in order not to be indignant about the situation. But *a reason to act* is not *a reason to feel*. The point is perfectly general. In §1 I argued that there are no reasons to know and no reasons not to know. Surely, it might be objected, there are often extremely good reasons not to know or discover that *p*. Knowledge is often painful, hard to bear, unpleasant. When this is the case, one has a reason not to do whatever might lead to knowledge.

3 Motives

Reasons, I said, are reasons *to* or not *to* believe, judge, emote, desire, act. Such reasons are consistent with no-one believing, emoting, desiring or acting. They relate facts and possible mental episodes, states and actions. When a person does judge, believe, emote or act, we sometimes say that her reason *for* judging, believing, emoting, desiring acting was this or that. The switch from "reason *to*" to "reason *for*" goes hand in hand with a switch in what we call a reason. A reason *to F*, I suggested, is a fact. But someone's reason *for F*-ing is *either* a fact – a true proposition or an obtaining state of affairs – *or* a proposition. Someone's reason for *F*-ing is either a reason to *F*, a fact, on the basis of which that person *Fs* or the propositional counterpart of that fact, on the basis of which that person *Fs*. The propositional counterpart of the fact that *p* is the proposition that *p*. A reason or its propositional counterpart on the basis of which someone believes, emotes, desires, acts is a *motive*. We speak of motives *for* believing etc. as we speak of reasons *for* believing etc.

The distinction between facts and their propositional counterparts does not in itself indicate any relation of priority between the two in the roles for them just suggested. Such a relation of priority may be introduced by saying that someone's reason for *F*-ing is either a reason to *F*, a fact, on the basis of which that person *F*s or *merely* the propositional counterpart of that fact, on the basis of which that person *F*s.

How are "on the basis of" and "for" to be understood? In terms of "because". Two very different types of because may be distinguished – the personal because and the impersonal because. Examples of the latter include

> It is true that *p* because the state of affairs that *p* obtains
> Sam died because the bullet went through his brain

Examples of the personal because include

> Sally slapped Sam because he is a sexist
> Sally slapped Sam because she believes him to be a sexist
> Sally slapped Sam because she despises sexists

(The distinction between the personal and the impersonal because is a counterpart of such familiar distinctions as those between personal and impersonal uses or senses of "refer", "mean" and "explain"). The first of these last, three examples employs *the personal because of objective reasons*: Sally's reason for slapping Sam, her motive, is the fact that he is a sexist. The second and third examples employ the *personal because of subjective reasons*. In each case, Sally's reason for slapping Sam is a fact, but a psychological fact involving Sally, her belief in one case and her scorn in the other. In each case, we are told that her reason for slapping Sam is that he is a sexist but not that it is a fact that he is a sexist nor that Sally slapped him because he is a sexist. From

> It is true that *p* because the state of affairs that *p* obtains
> Sam's reason *for* emoting/judging/desiring/acting was that *p*

we cannot infer that *p*. From all instances of

> *p* because *q*
> we may infer that *p* and that *q*.

The peculiarities of the personal because emerge very clearly in one type of self-ascription which employs a relative of the personal because, "for". Consider

> I suppose he was very conceited, for he was much addicted to judging his age (Henry James, *The Bostonians*, Bk 2, ch 21)
> He may die, for he has a terrible fever
> Come here! – for I have something very important to tell you.

"For", like French "car", is preceded by a comma. (The German "Denn", as Severin Schröder pointed out to me in Toledo, is written with a capital letter and preceded by a full stop, whereas the German word for "because", "weil", is preceded by a comma). "For" here is followed by a specification by the speaker or narrator of her reason – for supposing, asserting and ordering.

If Sally slapped Sam because he is a sexist, what is her intentional relation to the fact that he is a sexist? Does the claim imply that she knew that he is a sexist, as is sometimes asserted? Or does the claim merely suggest or implicate that this is the case? Here we arrive finally at the first of the possible relations mentioned in the Introduction between intentionality, on the one hand, and the theory of reasons, on the other hand.

When we are told that Sally slapped Sam because he is a sexist, the relevant because might be taken to be an impersonal one, the causal because, a merely causal because. But if it is interpreted as a personal because, does the claim really imply that Sally slapped Sam because she knew that he is a sexist? If that were the case, then our example of a personal because of objective reason would imply a personal because of subjective reasons, albeit one which is unusual in implying a psychological relation to a fact.

Some progress in understanding the relations between motives and reasons and between subjective and objective motives may perhaps be made by looking at a type of mental episode which contemporary theories of intentionality often ignore.

4 Operations & Correctness

What sort of things enjoy the property of intentionality? Attitudes, such as admiration and belief, make up one species of mental state which enjoys the property of intentionality. Attitudes are either thatish, like belief-that and regret, or not, like admiration, belief in and believing someone. The class of attitudes is often assumed to be exclusively thatish and almost equally often to consist of attitudes towards propositions or states of affairs. But properly understood, attitudes comprise both thatish and non-thatish cases and their distinguishing feature is that they either admit of degrees (anger, certainty, sadness, fear, desire, surprise) or have polar opposites (belief-disbelief, certain-uncertain, sadness-joy, like-dislike, desire-aversion).

Knowledge, perception and memory then constitute a distinct, *second*, kind of mental state or episode enjoying intentionality. For they admit of no degrees and have no polar opposites. Knowledge is not any sort of attitude. Nor are perception and memory.

A *third* family comprises decisions, intentions and judging which display neither degrees nor polar opposites. A *fourth* family of mental episodes is made up of what have been called mental operations or "synthetic" acts. Consider, for example: conditional judging, judging that *q* within the scope of the supposition that *p*; identification, as when one identifies an object one is observing at *t* with the object one observes a second later; distinguishing, as when one distinguishes perceptually one object from another; conjoining assertions or judgings. Just as conditional judging is to be distinguished from judging that (if *p*, then *q*), so too, identifying is to be distinguished from a judgment of identity and distinguishing from a judgment of numerical difference. In the same way, one can distinguish between the operation of denying that *Fa* and judging that it is not the case that *Fa* (or judging that *a* is not *F*).

These different examples of operations certainly involve mental acts and episodes which enjoy the property of intentionality – perception, supposing, conditional judging. But do the operations themselves enjoy intentionality? If Sam hypothetically judges that Sally is to be avoided under the supposition that she despises sexists, he certainly stands in an intentional relation to Sally and sexists and perhaps even to the relations of avoidance and scorn. He enjoys presentations or representations thereof. But does he also stand in an intentional relation to conditionality? Many philosophers have denied that even if Sam judges that (if Sally despises sexists, then she is to be avoided), he stands in an intentional relation to, enjoys a representation of, conditionality. Philosophers who deny that the logical constants represent anything should presumably reject the view that to disjoin, to identify, to distinguish and to deny involve representations of disjunction, identity, difference and negation. But even the logical realist about disjunction, conditionality should reject the view that disjoining and conditional judging represent anything. For operations have no sense and express no concepts. Synthetic acts, then, seem to be connections between acts or states which enjoy the property of intentionality but do not themselves enjoy this property.

The relations between the four types of mental episodes we have distinguished emerge more clearly if we consider their relation to the theory of correctness conditions. Believing that *p* and disbelieving that *p*, like certainty that *p*, are correct or incorrect. Are non-intellectual attitudes also correct or incorrect. If we follow Plato and Aristotle, and those who have followed them, such as Brentano and his pupils, we shall say that admiration, fear, joy, sadness, indignation and desire are all correct or incorrect. Just what their correctness conditions are is much more controversial than the correctness conditions of believing. But the general idea may be illustrated by claims such as

It is correct to pro/contra emote that *p* iff It is good/bad that *p*
Fear of the dog is correct iff the dog is dangerous
Admiration of *x* is correct iff *x* is admirable (elegant, wise...)
Indignation that p is correct iff it is unjust that p
Shame that *p* is correct iff it is shameful that *p*
Grief about the fact that *p* is correct iff the fact that *p* constitutes a loss
Love of love is correct iff love is valuable
x's desire to *F* is correct iff *x* may/ought to *F*

Our third category of mental states and episodes also enjoys correctness conditions:

Judging that *p* is correct iff *p*
x's intending to *F* is correct iff *x* may/ought to *F*

Knowledge, perception and memory, on the other hand, have no correctness conditions. The point is a generalisation of Aristotle's remark that (scientific) knowledge cannot be correct. It is true that to see a dog as an *N* has correctness conditions. To see a dog as an *N* is correct only if being an *N* is not incompatible with being a dog. To see a dog as a cat is to misperceive it. But seeing a dog is neither correct nor incorrect. It is a relation to a dog.

Operations, like attitudes and the judging, intending, deciding, conjecturing family, have correctness conditions:

It is correct to disjoin *p*, *q* iff *p* or *q*
It is correct to conjoin *p*, *q* iff *p* and *q*
It is correct to deny that *p* iff not-p

It is correct to identify x, y iff x = y
It is correct to distinguish x, y iff Not (x =y)

5 Some Relations between Reasons, Motives & Intentionality

We are now in a position to better understand some of the relations between intentionality, reasons and motives.

5.1 The Semantic Value of Personal Because

What do ascriptions of motives ascribe? We say that someone believes, emotes, desires, acts because *p*, where the fact that *p* is either a psychological fact – the personal because of subjective reasons for believing etc. - or a non-psychological fact – the personal because of objective reasons for believing etc.. What is the semantic value of the "because" in an ascription of motives? The question resembles the much easier question: what is the semantic value of "distinguishes" in an ascription of an act of distinguishing, as when we say of Sam that he distinguished Sally from Maria in the crowd? The answer to the latter question is that the ascription ascribes an operation, the operation of distinguishing. Such an ascription, we suggested, is not be confused with the attribution to Sam of a perceptual judgment to the effect that Sally is not Maria. If we take the analogy seriously, we should say that to say that Sally slapped Sam because he is a sexist is to ascribe to Sally an operation of connecting an action and a fact, perhaps a fact she knows, in a certain way. Similarly, we should say that to say that Sally slapped Sam because she believes him to be a sexist is to ascribe to Sally an operation of connecting a belief and what she believes and an action.

But is there really any such operation corresponding to certain uses of "because" and "on the basis of"? There is a radical alternative to the affirmative answer to this question. It is an answer suggested by some things Wittgenstein says. Someone believes, emotes, desires because *p* iff that person would or could reply to the question "Why do you believe, admire, desire...? with the answer "Because *p*".

Someone who is capable of answering such a question in this way understands the concept expressed by "because". But does the ability to believe, emote, desire act on the basis of the fact that *p* really presuppose a grasp of the meaning of "because"? Consider the parallel cases of denying, distinguishing and conditional judging. Children learn to use "No" before they learn to use "not". It seems equally plausible to think that a mastery of conditional judging precedes a mastery of judging of conditional contents, that a mastery of the operation of distinguishing, for example, perceptually, objects precedes the mastery of the concept of numerical difference.

Wittgenstein's suggestion at §487 of the *Investigations* that the origin or source of our knowledge of why we do what we do is a *feeling*, for example, that we have carried out some action because we have been ordered to do so, is certainly correct. When we distinguish two objects in our visual field or perform a conditional judging we do not feel either difference or conditionality. Nor do we feel any operation. But unless one is a friend of some extreme form of phenomenalism or sensualism it does not follow that there is no connection (Wittgenstein's word is *Zusammenhang*) between what we do and our motives other than that brought into being by our answers to "Why?" questions.

5.2 Correctness Conditions, Correctness Makers & Non-Defeasible Reasons

The claim that attitudes and mental operations as well as some other mental acts and states have correctness conditions can be strengthened in a readily understandable way. If the danger of a feared object, for example a dog, figures in the correctness condition of fear, it is natural to say that the danger of the dog makes fear of it correct. Such a fear is, then, not an ungrounded fear. Fear of a dog which is not in fact dangerous is, then, ungrounded. Similarly, the shamefulness of a deed makes shame correct and the fact that *p* makes judging and believing that *p* correct. This way of strengthening correctness conditions may seem to be particularly attractive to friends of truth-making. But although correctness-making and truth-making are indeed attractive to a certain sort of realist, the theory of correctness-making is arguably independent of the theory of truth-making.

Correctness and incorrectness are contradictories, truth and falsity, on many accounts, are contraries. What is true or false is not what is correct or incorrect. The bearers of correctness and incorrectness are states of believing, acts of judging, fear, indignation, desires. Where truth and falsity have bearers, these are propositions or whatever it is which is specified by that-clauses. It is true that we often say of the belief or judgment that *p* that it is true but we do not say this of believing or judging that *p*. Finally, truth is not any sort of normative property whereas, on one popular view, correctness is.

One way of understanding correctness in normative terms is as follows:

x correctly *F*s iff *Fx* and Ought (*Fx*)

The deontic concept employed here is the *ought to be* connective and not any sort of *ought to do*. Fuller versions of the general idea can be given for different types of acts and states. For example,

> *x* correctly believes that *p* iff *x* believes that *p* & *p* & Ought (if *x* believes that *p*, then *p*)
> *x* is right to be indignant that *p* iff *x* is indignant that *p* & It is unjust that p & Ought (if *x* is indignant that *p*, then it is unjust that *p*)

One alternative to this sort of account of correctness is the view that correctness is a *sui generis* normative property, another that it is not in fact a normative property.

Not only are truth and correctness very different properties, correctness-making is arguably a different type of making than the making peculiar to truth-making. If correctness is a normative notion, then correctness-making is normative grounding. Truth-making, on the other hand, is just metaphysical grounding. If that is right, then the internality of the reason relation differs fundamentally from the internality of the relations between numbers and colours. But to pursue this suggestion here would take us far away from our topic.

What is important here is the claim that every correctness-maker for attitudes and judging, intending, desiring and preference constitutes a non-defeasible reason for the relevant mental state or acts. The notion of a non-defeasible reason, then, should be understood in terms of correctness and correctness-making.

5.3 Reasonableness – the Ideal Case

What is it to be fully reasonable? It is to judge, believe, emote, desire, prefer, act on the basis of what makes judging, emoting, desiring, acting correct and to know that these correctness-makers obtain. Fear of a dog which is based on its danger, a danger known to the fearer, is as reasonable as fear can hope to be. Similarly, belief that *p* which is based on the fact that *p* and which is such that the believer knows that *p* is as reasonable as belief can hope to be.

Many philosophers find the former claim, about fear, much more plausible than the latter claim, about belief. For many philosophies of knowledge take knowledge to be a form of belief. But our taxonomy of mental episodes and states in § 4 gives one reason for thinking that belief-first accounts of knowledge are profoundly mistaken.

We suggested in § 4 that attitudes are mental states which either admit of degrees or have polar opposites. It is for this reason that all attitudes are *reac-*

tions or *responses*. That emoting is a reaction is widely accepted. But belief, certainty and conviction are also reactions or responses, precisely because they admit of degrees (certainty, conviction) or have polar opposites (belief-disbelief). What, then, are they reactions to? The foregoing suggests that attitudes are responses or reactions to what is known or to what seems to be known.

This view is compatible with the view that knowledge-that is not any form of belief-that. But it is also in fact compatible with the view that knowledge-that is a form of belief-that. It requires only that knowledge be prior to belief. Knowledge-that is a state or disposition. Finding out that *p*, coming to know that *p*, discovering that *p*, on the other hand, are episodes. The state of knowledge is axiologically prior to epistemic episodes. We typically discover or find out whether *p* because we want to be in the state of knowing whether *p*. Knowledge-that is a desirable possession. But axiological priority is not ontological priority. Cognizing, apprehending, coming to know that *p* marks the beginning of the state or disposition of knowledge-that. Impressed by the undoubted axiological priority of knowledge-that over apprehending and discovery, epistemology has for a long time entirely neglected the ontological priority of coming to know that *p* (*Erkennen*). Once this ontological priority is accepted, nothing stands in the way of accepting the traditional belief-first view that knowledge-that is a qualified type of belief-that provided we bear in mind that coming to know that *p* is not any form of judging or occurrent belief. The knowledge-first view of apprehending and discovery is compatible with the belief-first view of the state or disposition of knowledge-that.

Of course, most of the time, our motives for judging, emoting etc. are reasons which are only non-defeasible reasons. Indeed, we often emote, believe and act in an unmotivated, passive, minimal or automatic way. So called non-intentional actions such as doodling and scratching one's cheek, the affective states which make up day-dreaming, the desires and affective states which are drive-driven often have a good claim to be entirely unmotivated. Whimsicality is everywhere. In between the two extremes, full reasonableness and whimsicality, there lies the enormous variety of defeasible reasons to and not to judge, believe, desire, prefer, emote, act. What, then, is the relation between non-defeasible reasons and defeasible reasons? Here I have attempted to present some of the building blocks for an answer to this question. An answer which takes all the defeasible reasons to be, e.g., indignant that *p* to be the facts which make it more probable that it is unjust that *p* than not. The hope is that in such a way the concept of a defeasible reason can be understood in terms of the concept of non-defeasible reasons. But this is not the place to explore the relations between correctness-making and probabilification.

Peter SIMONS

Heads Road Safety

Why Agency and Intentionality were Made for each other (and how this Refutes Idealism)

1 Pedestrian Road Safety in London, 2006–2010

I start not with philosophy but with some salutory empirical facts. In the year 2013, 23,066 road traffic collisions involving personal injury were reported to the Metropolitan and City of London Police. Of these, 132 involved fatalities, 2,192 serious injuries, the remaining 24,875 slight injuries. Around half of the fatalities (65) were pedestrians.[1] The vulnerability of pedestrians in traffic accidents is obvious. In 2012 a report commissioned by Transport for London was published by the Transport Research Laboratory with the title *Analysis of police collision files for pedestrian fatalities in London, 2006–2010*.[2] The report used STATS91, the national database of all reported injury accidents on public roads, recording around 50 variables for each accident, and the Archive of Police Fatal Road Traffic Collision Files, compiled using an extremely detailed standard format called a Haddon's matrix. A sample of 198 fatalities was studied. Among the many detailed analyses contained in the 144-page report are statistics concerning the causes of fatal accidents. Some standout facts from this analysis are

> 33% of pedestrians were over 70, 20% over 80
> 24% of pedestrians were impaired by alcohol and/or drugs, 38% of the age range 16–59
> 48% of pedestrians were recorded as having "failed to look properly"
> 12% of pedestrians were distracted by following others, talking, using a mobile phone, eating etc.
> The most commonly recorded contributory factor for vehicles was "failed to look properly", recorded for 20% of vehicles; this was most common for all vehicle types except for

1 Transport for London 2014.

2 Knowles *et al.* 2012. Transport for London (TfL) is a local government body responsible for the transport system in Greater London. TfL has responsibility for London's network of principal road routes, for various urban rail networks, for London's trams, buses and taxis, for cycling provision, and for river services. The Transport Research Laboratory is a private consultancy offering services to the public and private sectors. Established in 1933 by the UK government as the Road Research Laboratory, it was privatised in 1996.

https://doi.org/10.1515/9783110560282-029

HGVs (Heavy Goods Vehicles), for which "vision affected by blind spot" was more common (recorded for 12 out of 27 HGVs).

Among the measures recommended in the report for reducing casualties to pedestrians are education of pedestrians and drivers, enforcement of the law, especially with regard to vehicle speed and driving under the influence of alcohol, and engineering improvements to roads and vehicles. Of these, the most relevant for us concern education.

2 Road Safety Education

Between 1951 and 2006, an astonishing total of 309,144 people were killed and 17.6 million injured on Britain's roads. The lowest recorded annual total of fatalities, 1,713, was in 2013, when there were 30.9 million registered motor vehicles. By contrast, in 1934, when there were fewer than 2 million registered motor vehicles, there were 7,343 fatalities, and the peak was during wartime blackout in 1941, at 9,169 fatalities, again with reduced vehicle numbers on the roads due to fuel rationing. Casualties have since been steadily reduced, so that now in Europe, only Sweden's road safety is better than Britain's. The improvement is due to many factors, one of which is better road safety education, for drivers as well as pedestrians. When I was child, I was taught what was then called the *kerb drill*. This went as follows: If you want to cross the road, *halt* at the kerb. Look right, look left, look right again, and if the road is clear, *quick march* across the road. This simple sequence, drilled into children to make it an automatic habit, kept millions from harm. Notice that you look *right* first and last. That is because in Britain, as in Japan, and in many former parts of the British Empire such as Australia, India, South Africa, and Hong Kong, traffic drives on the left. In continental Europe and the Americas, the instructions would switch 'left' and 'right'. In London at pedestrian crossing points on busy junctions and roads, the legends 'LOOK RIGHT' and 'LOOK LEFT' are painted onto the road in front of the kerb. This is because London hosts millions of tourists from countries which drive on the right, who are either unused to being pedestrians or who habitually look the other (in Britain, the "wrong") way. It is notable that the highest rate of pedestrian fatalities in London is in the City of Westminster, where Buckingham Palace, the Houses of Parliament and Westminster Abbey attract a high density of tourists.

The simple safety I learnt as a child was successively revised and refined as road conditions changed. In 1970 the somewhat militaristic kerb drill was replaced by the Green Cross Code, with a costumed superhero, Green Cross Man, designed to make it appealing to modern children. The Green Cross Code is still in use. The 2005 version goes as follows:

> *THINK!* Find the safest place to cross, then stop.
> *STOP!* Stand on the pavement near the kerb.
> *USE YOUR EYES AND EARS!* Look all around for traffic, and listen.
> *WAIT UNTIL IT'S SAFE TO CROSS!* If traffic is coming, let it pass.
> *LOOK AND LISTEN!* When it's safe, walk straight across the road.
> *ARRIVE ALIVE!* Keep looking and listening

Improvements are ongoing: pedestrian-controlled crossings, speed bumps, lower speed limits, vehicles with daytime lighting, collision sensors and automatic braking devices are all helping to reduce casualties. It is now common to see small children in kindergarten groups with their teachers all wearing high-visibility vests, and in some cases tethered together in groups to a single rigid harness. Campaigns to make four-wheeled motorists more aware of motorcyclists, requirements for vehicles to use headlights in daytime, together with vastly improved safety protection in vehicles, traffic calming schemes, and much more, have all contributed. Television advertisements, graphically and realistically illustrating the potential fatal effects of phoning or texting while driving, are meant to shock drivers, especially younger ones, into awareness of the dangers of inattention at the wheel. Modern cars use software which alerts drivers who change lanes without signalling, to counteract lane drift due to tiredness or inattention, and some software will even automatically apply the brakes in some situations to prevent or ameliorate a collision.

3 Coordinating Agency and Intentionality

Much of this is applied common sense. The reason I have dwelt on it at length is that it is a salient and important instance where the coordination of perception and action are highlighted because the effects of not doing so can be dramatic: fatal or life-changing. Both the kerb drill and the Green Cross Code for children contain injunctions for both *perception* (Look! Listen!) and *action* (Stop! Wait! Walk quickly!), and, more crucially, their coordination. Recommendations and requirements such as wearing high visibility clothing, using lights and signals, providing mirrors and cameras to enhance all-round vision, use of sensors, avoiding or reducing impairment due to alcohol, drugs, fatigue and distractions are all aimed at improving the *perception* by road users of hazards and other road users, getting them to *act* accordingly, and so reducing harm.

What has been said about road safety, applies to almost all areas of life, serious as well as frivolous, although few areas are so potentially widely deleterious to life and welfare, which is why I chose that topic as a way in. One of the most important things we learn as we grow up is how to coordinate our limbs in order to achieve our aims, be they drinking and eating, avoiding harm, moving about, manipulating objects, interacting with others. From the first clumsy grasp of a fluffy toy or a piece of food and the first faltering steps, to the fine-tuned movements of a surgeon, artist, sportsperson or concert musician, huge parts of our lives are devoted and dedicated to this coordination, in which our perception and motor activity are intimately intertwined. In fact, the cases where our mental lives disengage from action, such as in musing, daydreaming, pondering, thinking (whether idly or determinedly), passively contemplating a view, or listening to music, are the marked exception to the unmarked rule.

4 What is the Philosophical Significance of This?

The philosophy of intentionality has in the past tended to concentrate unduly on this passive or contemplative side. Consider for example the case of Brentano, single-handedly responsible for the reintroduction of intentionality into the conceptual armoury of Western philosophy in his *Psychologie vom empirischen Standpunkt*. Brentano's initial Cartesian scepticism about the existence of a mind-independent world led to his framing the notion of intentional inexistence as one dividing phenomena. The 1874 distinction is not one between mental and physical *entities*, such as thoughts on the one hand and trees on the other, but an internal distinction among phenomena, between those having, and those lacking an object contained or inexisting within themselves. The former are physical phenomena; the latter are mental. This is reinforced by Brentano's examples. Here are the physical phenomena he lists:[3]

> A color, a figure, a landscape which I see, a chord which I hear, warmth, cold, odor which I sense; as well as similar images which appear in the imagination.

Despite the frantic attempts by Oskar Kraus in a footnote to the 1911 edition to excuse what he claims is a mere slip on Brentano's part in the case of the land-

3 Brentano 1995, 79 f.

scape "which I see", it is clear from *all* of the examples that physical phenomena are all *phenomena*, that is, appearances, and so appearances *to* someone, and therefore not independently existing physical entities. This defect was only gradually remedied in publications of the 1890s as Brentano's students Meinong, Höfler, Twardowski and Husserl took a realist line and made a distinction between the content and object of mental activities, the former being either mental or a universal with only mental instances, the latter being of any ontological category. Brentano himself also revised his views along similar lines, but in part because of his visual impairment this did not make its way into print until the 1911 second edition, when Kraus appended several later notes and dictations to the *Psychologie*, and liberally sprinkled the reprinted body of the work with excusatory footnotes attempting forlornly and polemically to assimilate the earlier work to the later, against those criticisms.

In the case of Twardowski and Meinong, the distinction between content and object was fixed to the rigid view that every intentional act has an object, only that in some cases the object does not exist. Husserl on the other hand adopted the commonsensical and correct position that in such cases there is a content but no object. Whereas in his *Logical Investigations* of 1900–01, Husserl combines this sensible position with a straightforward realism about both the physical world and the platonic realm of logic and mathematics, by 1913 this gave way to his methodologically motivated transcendental idealism, which turned on the implicit but mistaken assumption that since it was not essential to all outwardly directed mental acts that they have objects, it is possible that none do, so that idealism cannot be ruled out. Therefore, the methodologically safe position for a philosophy intent on providing firm Cartesian foundations for knowledge is one where the natural attitude, that of tending to affirm as a default the extra-mental existence of objects of such outwardly-directed intentional acts, is disengaged from science in order that philosophical attention may be focussed solely on the undeniable acts themselves and their essential parts and concomitants.

The upshot of Brentano's initial and Husserl's later Cartesianism about the physical world is to demote considerations of action and agency to an insignificant corner of their psychology and phenomenology. Brentano's third and final class of mental phenomena, the class of so-called phenomena of love and hate, comprises both emotive and conative attitudes, but it omits considerations of action. Action could be expected to come into discussion of intentionality in at least three ways. The first is in the area of will, when mental activity translates into physical action: smiling at someone, reaching for an apple and biting into it, changing walking direction to cross the road, steering the car into the drive, and so on. The second is motor response to perceived conditions: ducking below a branch when walking, looking up when hearing one's name called, stopping at the kerb to look before crossing a road, applying the car's brakes to avoid hitting a pedestrian who has walked carelessly out into the road. The final and often neglected area is the conscious and intentional monitoring of ongoing action, the intentionality that accompanies and aids the controlling of agency. This, like the forms of proprioception which are crucial to motor control, has tended to evade phenomenologists' purview. Only a few realist phenomenologists such as Johannes Daubert and Maurice Merlau-Ponty stressed the importance of the body and our perception of it in their work. A laudable later exception is David Woodruff Smith, whose 1992 paper 'Consciousness in Action' stands out against the general trend.[4] This is not to deny that the controlling of activity is in good part not a conscious matter. In the advanced and finely controlled actions of skilled sportspeople, dancers, and musicians, much of the work done by what is called 'muscle memory' is precisely not consciously controlled, enabling attention to be directed elsewhere.[5]

5 Essential Connections

Consider an ordinary everyday sequence of human actions. I am running short of food and want to get some for a meal I am making for friends in the evening. I write a list, exit the house and lock it, unlock and enter the car, start it, drive to the supermarket, locate, choose, take and pay for the items, load them into the car, drive it back home and unload my purchase. The compound sequence of actions is driven by the intention to provide a meal for friends that evening, consideration of the means for doing so, and knowledge how to use these means to that end. At each stage the action is motivated by that overall intention and guided by my knowledge of how to go about realizing it. As each subaction in the sequence unfolds, I am consciously as well as unconsciously monitoring and controlling how it goes, adjusting my simpler actions to the perceived circumstances such as the road conditions, the location of items in the supermarket, my budget and so on.

The actions are directed to an end, a desired state of affairs that I have consciously in mind. That is desiderative intentionality. My knowledge of how to

4 Smith 1992.
5 A surprising early proponent of this idea was the German astronomer Friedrich Bessel.

achieve the end is partly habit, partly memory, partly deliberation: again, conscious intentionality is involved. My monitoring the circumstances as they unfold is perceptual and cognitive intentionality and controlling the actions in fine detail is perceptual and proprioceptual intentionality as well as muscle memory. There is then no way to execute the action and realize the intention without conscious intentionality: desire, intention, memory, perception, judgement, monitoring. But equally, those intentional acts are subservient to the overall aim of satisfying the intention by doing the things that will realize it. The actions and the intentional acts are mutually supporting and mutually motivating. Action requires intentionality, intentionality serves action. There can be activity such as finger-tapping or head-scratching without intention, and intentionality without overt activity such as daydreaming or idle recalling. But in general (not without exceptions, but in the main), action without intentionality is aimless, while intentionality without action is pointless.

6 Why Agency and Conscious Intentionality in Fact Co-Occur

During the heyday of behaviourism, it was common for psychologists and their philosophical admirers to take the position that successful agency does not require consciousness, or that conscious states can and should be ignored in the scientific evaluation of the way the organism fits into and responds to its environment. Even a philosopher sympathetic to the occurrence of conscious acts might adopt the parsimonious position that consciousness as such and its associated intentionality are always or at least where possible to be kept out of causal and explanatory accounts of the feedback between environmental conditions and animal behaviour. Any philosopher considering like Davidson that consciousness depends on language would be bound to see the sensory–motor routines of non-linguistic animals as unmediated by intentionality.

Against this I want to suggest as highly plausible the view that intentionality and agency are mutually reinforcing advantages in animal evolution. A characteristic and universal mark of animality in the biosphere is motility, the ability to move (at some stage in the animal's life). Motility of the whole organism (locomotion) is clearly of evolutionary advantage, for securing food and mates, and for avoiding predators and other harms. The use of separately motile parts (limbs) likewise contributes not only to organism motility (see a starfish "walking", a bird flying or a fish swimming) but to the accomplishment of other tasks that aid survival, such as grasping, hitting, injuring, biting and signalling. However, all of these abilities can only be evolutionarily advantageous to the extent that the organism directs its movements appropriately. This is obviously where the sensory systems come in. Take the most important distal sensory organs, eyes. Whether eyes evolved once or several times, the advantages of a directional detector of distal light and shape were so obvious and manifold that the development of eyes is thought not only to have triggered an evolutionary arms race during the Cambrian period, but to have been responsible for the development of central nervous systems and brains. Nerves connected eyes with muscles, which means that the mechanism for detecting distal light and shapes co-evolved with mechanisms for motility, which had previously been largely confined to beating cilia.

Once the neural system was sufficiently complex to integrate stimuli from large numbers of nerves, and to control motility with some precision, the basis was laid for the representation of the state of parts of the nervous system, to other parts, which may well be the physical precondition for conscious awareness. The addition of other systems of distal perception – hearing, olfaction, sonar, infrared detection, electric field detection – would also enable animals to exploit different media of information to aid in food detection, avoidance of predators and other dangers. Consciousness may have evolved as a means for integrating the information from the various sensory modalities and allowing it to guide activity taken in response. From this point of view, there is every reason to expect that consciousness would be a rather widespread feature of animal life, present not only in diverse classes of vertebrates such as mammals and birds, but even in different phyla: it is fairly certain that octopi, from the phylum *Mollusca*, are conscious. As we are unable to get inside the minds of other creatures, this must remain conjectural at least for now, but the indications from similarities of neural functions and complexities of behaviour are rather compelling.

7 How These Facts Refute Idealism

I am understanding 'idealism' as an ontological doctrine, to the effect that all things are either themselves minds or spirits or are in some way ontologically dependent on the mental. Classical idealists include Leibniz and Berkeley, for whom God and other spirits are aspatial substances and everything else depends on them. This renders the natural world existentially dependent on minds: were there no minds, there would be no matter or other physical entities.

Transcendental idealists such as Kant and Husserl maintain that our epistemic and cognitive situation entails that we can have no knowledge (Kant) or no grounded knowledge (Husserl) regarding things as they are in themselves. This need not entail classical idealism – Kant famously declared his opposition to Berkeley's idealism and pronounced in favour of the extra-mental existence of things in themselves. Husserl's cautious version of transcendental idealism, despite some rather classical-sounding statements in *Ideas*, is held by his realist sympathisers to be fully compatible with the independent extra-mental existence of material and other non-mental objects.

Idealism is in many cases adequately and conclusively exposable as resulting from committing a species of fallacy that David Stove referred to as "The Gem", and derided as "the worst argument in the world":[6] the argument from the incontestable facts that we have no access to things except through sensibility and cognition to the conclusion that we have no access to them as they are in themselves. This, though it is not the sole reason why Kant is a transcendental idealist (it leaves out the antinomies for example) is certainly one reason why transcendental idealism has been considered convincing. There is a clear version of The Gem in Berkeley's *Principles*:[7]

> When we do our utmost to conceive the Existence of external Bodies, we are all the while only contemplating our own Ideas. But the Mind taking no notice of itself, is deluded to think it can and doth conceive Bodies existing unthought of or without the Mind; though at the same time they are apprehended by or exist in itself.

However, as Franklin, commenting on Stove, observes, while The Gem is fallacious, that does not of itself explain how the mind is able to access things beyond itself, or, as he puts it, "how we can see, given that we have eyes".[8] Franklin dismisses three attempts to avoid the issue: naïve realism (too obviously wrong that we are in direct contact); divine assistance; and inference to the best explanation. His argument against secularised versions of the divine assistance explanation claims that attempts to invoke evolution are circular:

> Obviously, to invoke evolutionary theory as itself a reason to believe in sense perception is circular, since the only reasons to believe in evolution rely on sensory data.[9]

6 Stove 1995.
7 Berkeley 1710, § 23.
8 Franklin 2002, 624.
9 Franklin 2002, 624.

This argument, while not quite a Gem, is unconvincing. Firstly, it is not true that the sole reasons we have to believe evolutionary theory to be true rely on sensory experience. There are theoretical reasons too, such as parsimony, the avoidance of supernatural explanations, and its ability to integrate many facts and areas of inquiry. Secondly, there are other reasons to believe in the existence of veridical sensory perception than just those deriving from acceptance of evolutionary theory, namely the self-givenness of the independence of objects through their experienced recalcitrance and their robustness of impingement upon us. As Whitehead once put it to William Hocking, emphasizing the bodily and tactile side of experience, "*being tackled at rugby: there is the Real.* Nobody who hasn't been knocked down has the slightest notion of what the Real is."[10]

There is indeed a certain circularity in using data gathered through sensory experience as evidence to support a theory whose consequences include explanations as to why such experience can be veridical and demonstrative of independently existing objects of intentionality. That would be damning only if the circularity were vicious. But in fact it is virtuous: through sensory experience which we naturally take to be of an independently existing reality, we bootstrap our way through sophisticated but amply warranted and corroborated theory to a satisfying explanation as to why such experience is indeed veridical. To suppose we cannot get beyond in this way is indeed to succumb to the following Gem: There is no sensory evidence of independence independent of sensory evidence: therefore, there is no sensory evidence of independence.

The existentially urgent imperatives of road safety offer a stark reminder of the perils of an excessive intellectualism which discounts or downplays the connection between cognition and action. Those who seriously and consistently think that steep cliffs, poisons, loaded guns and speeding motor vehicles are dependent on the mind are invited to ignore their independence and play idealist roulette. They will be dead not in the long but in the very short run.

The parsimonious, scientifically accurate and explanatory account of both the veridicality and the action-linked nature of sensory intentionality is that such intentionality linked to action is survival-enhancing and has evolved and flourished in the animal kingdom for just that reason. No doubt it is possible to live a long life and sincerely believe in idealism: indeed, I know of at least three colleagues from Trinity College Dublin, namely George Berkeley, Arthur Aston Luce and David Berman, who have managed just that feat. I personally cannot understand what it's like to be a sincere and consistent idealist and still take

10 Hocking 1961, 512; Kline, ed. 1989, 13.

care when crossing the road, but maybe that is my crude realist imaginative limitation. For those who think like me, and regard idealism as at best a distraction from and at worst a rejection of the serious aims of scientific philosophy, it is both straightforward and obvious that, whatever the fine-grained detail about the nature of intentionality and the nature of action, about which I have said very little here, veridical accuracy of intentionality and efficacy of action are mutually reinforcing characteristics that we exploit and strive to improve for our own survival and enhanced welfare.

8 References

Berkeley, G., 1710, *A Treatise Concerning the Principles of Human Knowledge*. Dublin, Printed by Aaron Rhames.

Brentano, F., 1995, *Psychology from an Empirical Standpoint*. London, Routledge.

Franklin, J., 2002, Stove's Discovery of the Worst Argument in the World. *Philosophy*, 77, 615–624.

Hocking, W. E., 1961, Whitehead as I Knew Him. *Journal of Philosophy*, 58, 505–516. (Reprinted in: G. L. Kline, ed., *Alfred North Whitehead. Essays on His Philosophy*. Lanham: University Press of America, 1989, 7–17).

Knowles, J., L. Smith, R. Cuerden and E. Delmonte, 2012, *Analysis of police collision files for pedestrian fatalities in London, 2006–2010*. Transport Research Laboratory, Published Project Report PPR260 prepared for Transport for London. Accessed at: <http://content.tfl.gov.uk/pedestrian-fatalities-in-london.pdf>

Transport for London 2014, *Surface Transport – Casualties in London 2013*. Fact Sheet. Prepared by J. Stordy. Accessed at: <http://content.tfl.gov.uk/casualties-in-greater-london-2013.pdf>

Smith, D. W., 1992, Consciousness in Action, *Synthese*, 90, 119–143.

Stove, D. C., 1995, Judge's Report on the Competition to find the Worst Argument in the World, in his *Cricket Versus Republicanism*. Sydney, Quakers Hill Press, 66–67.

Wolfgang Künne

Intentionality in Husserl's *Logical Investigations*

A Plea for a (Minor?) Revision

I can't hang him when he doesn't exist; but
I can look for him when he doesn't exist.
Wittgenstein

1 Introduction

In the first part of this paper I shall discuss an important feature of Edmund Husserl's views on non-propositional thinking, thinking of an object. In the second part I shall examine an important ingredient of his account of propositional thinking, thinking that things are thus and so, and it is in this part that I shall plea for a revision of his conception of intentionality. As you will see in due course, I shall propose to *russell* Husserl at a certain point.

2 Non-Propositional Thinking

My point of departure is a passage in section 11 of Husserl's Fifth *Logical Investigation* that I take to be a key text as regards his conception of intentionality. Here are the first four sentences of this passage. (I shall read out the German text as well as J. N. Findlay's translation. As for the latter, I offer my condolences to all translators of Husserl. Unfortunately, no good translation into German is available either.)

Huss-1).[1]

[1] Ich stelle den Gott Jupiter vor, das heißt, ich habe ein gewisses Vorstellungserlebnis, in meinem Bewußtsein vollzieht sich das *den-Gott-Jupiter-Vorstelle.*
I have an idea of the God Jupiter: this means that I have a certain presentative experience, the *presentation-of-the-god-Jupiter* is realized in my consciousness.

1 Husserl LU, Vol. II/1, 373, my underlinings; Husserl LI, Vol. 2, 99.

https://doi.org/10.1515/9783110560282-041

[2] Man mag dieses intentionale Erlebnis in deskriptiver Analyse zergliedern, wie man will, so etwas wie de[n] Gott Jupiter kann man darin natürlich nicht finden; ... er ist also ... nicht immanent oder mental.
This intentional experience may be dismembered as one chooses in descriptive analysis but the god Jupiter naturally will not be found in it; ... hence ... he is not immanent or mental.

[3] Er ist freilich auch nicht extra mentem, er ist überhaupt nicht.
But he also does not exist extramentally, he does not exist at all.

[4] Aber das hindert nicht, daß jenes *den-Gott-Jupiter-Vorstellen* wirklich ist, ein so geartetes Erlebnis, eine so bestimmte Weise des Zumuteseins, daß, wer es in sich erfährt, mit Recht sagen kann, er stelle sich jenen mythischen Götterkönig vor, von dem dies und jenes gefabelt wurde ... [*Forts. folgt.*]
This does not prevent our *idea-of-the-god-Jupiter* from being actual, a particular sort of experience or particular mode of mindednes, such that he who experiences it may rightly say that the mythical king of the gods is present to him, concerning whom there are such an such stories... [tbc].

Let us try to decipher this fairly contorted piece of prose, and let us begin at the beginning. Does the first sentence in [1],

(s) Ich stelle den Gott Jupiter vor

really mean what it is said to mean in Husserl's statement [1]? Sentence (*s*) could be used by an actor who wants to tell us which role he plays in a performance of Kleist's *Amphitryon.* So let us insert a pronoun in order to obtain a sentence that comes closer to what Husserl wants to convey:

(*s+*) *Ich stelle mir den Gott Jupiter vor.*

The reflexive pronoun is no more *logically* reflexive than the pronoun in '*sich irren*', '*se tromper*', '*confundirse*'. Actually, in [4] Husserl himself uses the 3rd person variant of (*s*+). I suspect that it is not an accident that Husserl uses the odd (small-s) rather than (s+), for one tends to hear (s-plus) as a self-ascription of an act of visualizing an Olympian God, but Husserl may want something more neutral. Perhaps

(S) *Ich denke an den Gott Jupiter.*
I am thinking of the god Jupiter.

would better serve his purposes, for it leaves open whether I picture Jupiter to myself. As we shall see, a few lines after [4] in our passage, Husserl's unenviable translator renders (small-s) neither by 'I have an idea of Jupiter', as in [1], nor by 'Jupiter is present to me', as in [4], but by 'I think of Jupiter'. (I use the progressive because I find it more colloquial.)

In statement [2] of our passage Husserl rejects the idea that thinking of Jupiter consists in getting into a cognitive relation to something *intra*-mental, that is, he rejects the conception of intentionality that his teacher Brentano adopted, or apparently adopted, in 1874. In statement [3] Husserl dismisses the Twardowski-Meinong view according to which thinking of Jupiter consists in getting into a cognitive relation to something *extra*-mental.[2] In statement [4] he begins to present his own view according to which thinking of Jupiter does not consist in getting into a relation to *anything.*[3]

Standing-in-front-of really *is* a relation, and because of Jupiter's non-existence the two-place predication 'I stood 10 m in front of Jupiter and took a photograph of him' does not express a truth. After all, as Husserl maintains, "if a relation obtains the *relata* have got to exist, too (*besteht eine Relation, so müssen auch die Relationsglieder existieren*)."[4] Nevertheless, the sentence 'I am thinking of Jupiter' can very well be used to say something true. So we need an interpretation of (S) that explains this glaring difference. At this point we should heed Husserl's hyphenations that I emphasized in [1] and [4]. The hyphens in the sentence 'In my consciousness an act of thinking-of-the-god-Jupiter takes place' as well as Husserl's talk of a "mode (*Weise*)" in [4], are attempts at guarding against the impression that in asserting (S) one states that a certain relation obtains between oneself and an Olympian God.

Unfortunately, classifying (S) as self-ascription of a "*Weise des Zumuteseins*" is badly misclassifying it. Consider a standard use of the phrase '*mir ist so und so zumute*', and let Goethe set the standard. In *Dichtung und Wahrheit* (III.10) the poet reports his final bidding adieu to his mistress Friederike: "*Als ich ihr die Hand noch vom Pferde reichte, standen ihr die Tränen in den Augen, und mir war sehr übel zumute.*" Whatever Findlay may mean by 'modes of mindedness', *Weisen des Zumuteseins* are feeling uncomfortable, feeling wretched, feeling creepy, feeling comfortable, feeling fine, being glad, being overjoyed. If you ask somebody '*Wie war dir in dem Augenblick zumute?*' he cannot sensibly answer, 'At that moment, I was thinking of Jupiter (or of Bismarck)'.

---

2 Cp. Meinong 1904; Twardowski 1982 (1894).

3 Husserl suspects that here one can hardly avoid talking of a relation ('*die Rede von einer Beziehung* [*wird sich*] *hier nicht vermeiden lassen*'): Husserl *LU* II/1 (same section) 372.

4 Husserl 1979, 464. Cp. Husserl 1990-91 (1894), 150: '*von jeder echten Relationswahrheit* [*gilt:*] *das Sein der Relation schließt das Sein der Relationsglieder ein* (if a truth is really relational the being (obtaining) of the relation involves the being (existence) of the terms of the relation).'

Is Husserl's use of hyphens meant to suggest that we should regard the phrase 'thinking of the god Jupiter' as a seamless whole, as semantically unbreakable?[5] If that were so then (S) would no more entail 'I am thinking' than 'I am a stalker' entails 'I am a talker', – in other words, if 'thinking of the god Jupiter' were semantically atomic then the presence of 'thinking' in that phrase would be as much of an orthographic accident as the presence of 'talker' in 'stalker'. Furthermore, the phrases 'thinking of the god Jupiter' and 'thinking of the goddess Diana' would have no more in common (semantically) than the words 'violin' and 'violence'. But while understanding one of the words 'violin' and 'violence' does not help with understanding the other, in the case of 'thinking of the god Jupiter' and 'thinking of the goddess Diana' you have learnt part of the lesson required for understanding either of them as soon as you have come to understand the other. So we'd better not regard Husserl's hyphenations as suggesting semantical atomicity.

An interpretation of (S) that respects Husserl's strictures must meet two conditions: Firstly, (S) can express a truth even though the referent of 'I' does not stand in a relation to an Olympian God, and, secondly, (S) implies 'I am thinking'.[6] A bit of notation will help me to save ink when formulating my proposal. I use (Quinean) square brackets as a singular-term forming operator that transforms an expression into a designator of its sense. Thus the singular term '[superfluous]', say, designates the sense of the expression between the brackets. This operator can be used for disambiguating the sentence

(X) The sense of 'redundant' is superfluous.

If (X) is supposed to express a truth then it does not declare a certain sense to be superfluous, but rather it *identifies* that sense:

(Y) The sense of 'redundant' is [superfluous].

While in (X), under the unwelcome reading, 'is' functions as copula, the 'is' in (Y) is the identity operator. Now I think that in the language of classical predicate logic enlarged by the square brackets operator we can render (S) as follows:

---

5 This account of sentences like (S) was ascribed to the late 'reist' Brentano in Arnauld 1975, 197.

6 Cp. the discussion of Nelson Goodman's answers to the question *What is the logical form of sentences like 'This picture represents Pegasus'?* in section (2) of the appendix "*Fiktive Gegenstände*" in Künne 1983 (2007).

(σ) ∃x (x is an act of thinking & I am the agent of x & the content of x is [the god Jupiter]).

In uttering (σ) I express a truth just in case I perform an act of thinking the content of which is the sense that the phrase 'the god Jupiter' has in my utterance. The term 'content' is another word for what Bolzano called the *Stoff* of an intentional act or state and what Bolzano's admirer Husserl, latinizing '*Stoff*', called the '*Materie*' (or '*Auffassungssinn*') of an intentional act or state. My two requirements on an interpretation of (S) are fulfilled. Firstly, in uttering (σ) I do not ascribe to myself a relation to the god Jupiter but rather a relation to the sense of the phrase 'the god Jupiter': I say of a certain concept that it is the content of my thinking. The truth-value of (σ) does not depend on whether anything falls under this concept, that is, whether there is any Olympian God. (At this point I am pleased if the name of my brackets operator makes Husserl scholars hear a faint echo of the Husserlian operation of bracketing.) Secondly, (Sigma) entails '∃x (x is an act of thinking & I am the agent of x)', which is a 'Loglish' equivalent of 'I am thinking'.

Now if I am on the right track with (S) then we should capture the message of

(T) I am thinking of the Imperial Chancellor Bismarck

along the same lines. 'The mere fact of real existence', Elizabeth Anscombe once said, 'can't make so much difference to the analysis of a sentence like "X thought of —".[7] That seems dead-right. So I suggest this Loglish rendering of (T):

(τ) ∃x (x is an act of thinking & I am the agent of x & the content of x is [the Imperial Chancellor Bismarck]).

Just as in the case of asserting (σ), when I assert (τ) I do not ascribe to myself a relation to an object that is designated by the singular term that stands between the square brackets. In contrast to the case of (σ), this time such a relation does obtain, but in asserting (τ) we do not *state* that it obtains. If we understand (S) and (T) along the lines I drew then their truth-conditions are entirely analogous, and this dovetails with Miss Anscombe's semantic intuition. What is more important for my talk is that it complies with Husserl's theory of intentionality – as we shall now see in the continuation of my key passage. At the point at which I interrupted him Husserl goes on to say:

7 Anscombe 1981 (1965), 5.

Huss-1 (continued).

[5] Existiert andererseits der intendierte Gegenstand, so braucht in phänomenologischer Hinsicht nichts geändert zu sein.
If, however, the intended object exists, nothing becomes phenomenologically different.

[6] Für das Bewußtsein ist das Gegebene ein wesentlich Gleiches, ob der vorgestellte Gegenstand existiert, oder ob er fingiert und vielleicht gar widersinnig ist.
It makes no essential difference to an object presented and given to consciousness [¿?] whether it exists, or is fictitious, or is perhaps completely absurd.

[7] Jupiter stelle ich nicht anders vor als Bismarck, den Babylonischen Turm nicht anders als den Kölner Dom, ein regelmäßiges Tausendeck nicht anders als einen regelmäßigen Tausendflächner.
I think of Jupiter, as I think of Bismarck, of the Tower of Babel as I think of the Cologne cathedral, of a regular thousand-sided polygon as of a regular thousand-faced solid.

Unlike Findlay, I think that in statement [6] Husserl means by 'what is given to consciousness (*das für das Bewußtsein Gegebene*)' the intentional *act*, that is, in the case (S) and (T) an act of non-propositional thinking. It makes little, or no sense, to say that a certain German statesman, a certain Olympian God and a certain number are "for my consciousness essentially the same". But it does make some sense to say that no matter whether the object we are thinking of is Bismarck or Jupiter, the smallest or the largest prime number, we are always performing a mental act of the same kind. I take it that the "essential equality (*wesentliche Gleichheit*)" is captured by rendering all self-ascriptions of such acts in the style of (σ) and (τ).

The final pair of examples in statement [7] shows that we were right in not restricting the extension of '*etwas vorstellen*' to that of 'picturing something to oneself'. As Descartes pointed out, our *facultas imaginandi* does not enable us to distinguish a 1.000-sided polygon from polygons that are 999-sided or 1.001-sided, and I dare say that even Magritte is unable to picture to himself a regular thousand-faced solid. (What every mathematician knows – remember that Husserl started his academic career as a mathematician – is not clear to every philosophy student. Some of my students, at least, got confused because in the last pair of examples in [7] Husserl suddenly puts the *empty* term last. There are only five regular polygons, and none of them has more than 20 faces, as was demonstrated by the mathematician after whom Plato named his dialogue *Theaetetus*.)

According to Husserl's statement [5], a difference between thinking of X and thinking of Y that remains invisible if one practices *epoché* with respect to the question whether there is such a thing as X or Y is phenomenologically irrelevant.[8] Even the truth-value of reports of non-propositional thinking remains unaffected by the answer to that question. What does *not* become invisible by practicing *epoché* with respect to the question of existence is a difference in *content* between thinking of X and thinking of Y, so *this* difference is *not* phenomenologically irrelevant. And of course, it is also relevant to the truth-value of reports of non-propositional thinking. 'I am thinking of the author of *Candide*' may express a truth in Evita's mouth while 'I am thinking of the most famous philosopher at the court of Frederick the Great' does not, even though the author of *Candide* was the most famous philosopher at Sanssouci. (It is also possible that both reports are true in Evita's mouth although she has no inkling that one and the same person falls under the concepts that are the contents of her thinking.)

In statement [5] Husserl talks about the "intended objects" of the acts reported in (S) and (T), and from here it is a small step to saying that in (S) the intended object is a "merely intentional" object while in (T) it is an "existent" object. I think clarity is served if we refrain from putting things this way, for this manner of talking suggests that there are two kinds of objects, Jupiter belonging to the former kind and Bismarck belonging to the latter. I prefer to say that some acts of non-propositional thinking have objects, some don't. In cases like (T) there is something such that it is the intentional object of the act, in cases like (S) nothing is such that it is the intentional object of the act.

This was not a plea for a revision of Husserl's early theory of intentionality, for in *LI* Husserl himself maintains:[9]

> Huss-2)
> Der Gegenstand ist ein "bloß intentionaler" ... heißt: die Intention, das einen so beschaffenen Gegenstand 'Meinen' existiert, aber nicht der Gegenstand. Existiert andererseits der intentionale Gegenstand, so existiert nicht bloß die Intention, das Meinen, sondern auch das Gemeinte.
> The object is "merely intentional" ... means that the intention, the reference to a certain object, exists, but not ... the object itself. If the intentional object exists, the intention, the reference, does not exist alone, but the thing referred to exists also.

There is no proper subset of objects that contains all and only those that are "merely intentional", but there is a proper subset of objects – of existent ob-

---

8 The question whether a reader of Plato's *Critias* is thinking of Atlantis is entirely independent of the (still unsettled) question whether the island of that name was just a product of Plato's fancy or not.

9 Husserl *LU* II/1 (V, § 21) 425; *LI* 2, 127, slightly altered.

jects, that is – that contains all and only those that are at some time or other "intentional" objects. I regard being an intentional object to be a property like being floodlighted: it is a property that objects can acquire and lose, and it is a property that some objects never ever acquire. An object acquires the property of being an intentional object whenever it happens to be focussed upon by some subject of intentional acts or states, or to use Husserl's terminology or that of his translator, whenever it is "intended" or "referred to".

If you were so unkind to ask me for an example that belongs to the class of objects that never acquire the status of an intentional object I would stubbornly refrain from complying with your request. For if I were to answer, "Well, for example, the smallest pebble on the bottom of the Caspian Sea", I would fall into a trap, for by giving this reply I would have *made* that object an *intentional* object. This observation should not worry any realist. Compare this: If we stipulate that only names are admissible substitutes for the variable 'x' then the fact that there is no verifying substitution instance of '$\exists x$ (x is nameless)' does not prevent that objectual existential quantification from expressing a truth, and a fairly obvious one, even though every grammatically admissible substitution instance is self-refuting. Similarly, the exemplification resistance of the concept *Object that is not an intentional object* does not prevent that concept from being multiply instantiated.

3 Propositional Thinking

According to Husserl, propositional or 'thatty' mental acts and states do not only have a special kind of *contents* – I shall call their contents propositions –, they also have a special kind of intentional *objects*. Suppose

(E) Evita judges that Voltaire was witty.

According to Husserl, Evita cannot think *that* Voltaire was witty without thinking *of* Voltaire and *of* wit. These non-propositional acts of thinking are, Husserl would say, parts of her act of propositional thinking. Is there something that differs both from the man and from the property and that is the intentional object of the act of taking that man to have that property? Husserl's affirmative answer begins as follows:[10]

10 Husserl *LU* II/1 (V, § 17) 401–402; *LI* 2, 114, two misprints corrected.

Huss-3)

[a] Wie immer ein Akt aus Teilakten zusammengesetzt sein mag, ist es überhaupt Ein Akt, so hat er sein Korrelat in Einer Gegenständlichkeit. Und diese ist es, von welcher wir im vollen und primären Sinne aussagen, daß er sich auf sie beziehe.
Whatever the composition of an act out of partial acts may be, if it is to be an act [*better:* a single act] at all, it must have a single objective correlate, to which we say it is directed in the full, primary sense of the word.

[b] Auch die Teilakte ... beziehen sich auf Gegenstände; diese werden im allgemeinen nicht mit dem Gegenstand des ganzen Akte identisch sein, obschon sie es gelegentlich sein können. Natürlich kann man in gewisser Weise auch von dem ganzen Akt sagen, daß er sich auf diese Gegenstände beziehe, aber dies gilt doch nur in einem sekundären Sinn... (Forts. folgt.)
Its partial acts ... likewise point to objects, which will, in general, not be the same as the object of the whole act, though they may occasionally be the same. In a secondary sense, no doubt, the whole act may be said to refer to these objects also... (tbc)

The first type of examples Husserl has in mind is to be found in the realm of *non*-propositional intentional acts. If you are thinking of someone as the author of *Candide* then a person is the primary object of your thinking and a book is its secondary object. (Two *further* auxiliary objects come into play if you are thinking of Voltaire as the author of a book that makes fun of Leibniz's *Theodicée.*) Husserl forgot to give us an example that complies with his aside in [b] according to which *sometimes* the object of a subservient intentional act coincides with the object of the whole act. So let me make good for the omission. Suppose that my cousin Felix is the man whom Felix admires more than anybody else, and at this moment I am thinking of my cousin as the man whom Felix admires most. Then the embedded act has the same intentional object as the act it is embedded in. The same holds, *mutatis mutandis*, when I am thinking of the third natural number as the product of three and one.

But this was only an aside. It is the second type of complex acts Husserl has in mind that I want to discuss. I take the liberty of inserting my own example into the text:

Huss-3 (continued).

[c] *[Wenn jemand urteilt, dass Voltaire witzig war, dann ist Voltaire] zwar der Gegen-stand, "über" den geurteilt wird ...; aber gleichwohl ist [er] nicht der primäre Gegenstand, nicht der volle des Urteils, sondern nur der des Urteilssubjekts. Dem ganzen Urteil entspricht als voller und ganzer Gegenstand der geurteilte* Sachverhalt...
[If we judge that Voltaire was witty then Voltaire] is the object *about* which we judge... [He] is not, however, the primary or full object of the judgement but only the object of its subject [sc. act]. The full and entire object corresponding to the whole judgement is the *state of affairs* judged...

[d] Und dieser Sachverhalt ist offenbar nicht zu verwechseln mit dem bezüglichen Urteil.
This state of affairs must obviously not to be confused with the judging of it.

There is a rather unfortunate shift of meaning of the term '*Urteil*' when Husserl moves from [c] to [d]. In [c] judgements figure, as was to be expected, as examples of complex intentional acts. But the misidentification Husserl wants to avert in [d] is *not*, as Findlay quite understandably presumes, that of mistaking the state of affairs that Voltaire was witty for the act of judging that he was. It is extremely unlikely that anybody is prone to that mistake. As the continuation of the passage shows, the temptation Husserl wants us to resist is rather that of taking that state of affairs for the *content* of this act.[11] For various reasons Bolzano knew, '*Urteil*' is a bad name for this content. So let us stick to 'proposition'.

Why should one assign to propositional acts a special sort of objects at all? Husserl's answer is to be found in the first statement in part [a] of (Huss-3): The unity of a complex act, he maintains, depends on its having its own objective correlate. For two reasons I find this argument rather weak. First of all, Husserl himself concedes that some complex *non*-propositional acts have *no* objective correlate without thereby suffering from lack of unity: my thinking of the notoriously unfaithful husband of the goddess Juno, for example, is one complex act but this act does not owe its unity to its having an object, so the unity of complex acts is certainly not *always* due to their having their own objective correlate. Secondly why should the unity of a complex intentional act of whatever kind not be provided by the unity of its *content*?

But let that pass. Let us accept Husserl's ill-argued contention that Evita's judgement that Voltaire was witty does have its own objective correlate, namely the state of affairs that Voltaire was witty. Now if we want to single out the *content* of her act we will use the very same that-clause. But the content of a mental act or state is never identical with its object. So that-clauses are systematically ambiguous.[12] States of affairs are more coarsely individuated than propositions. If John says to Carmen, 'You are Spanish', and Carmen concurs, 'Yes, I am', then they give voice to two acts of propositional thinking that are, as Husserl would put it, directed at the same state of affairs: "the same state of affairs is grasped, is conceived of, in two different ways (*derselbe Sachverhalt wird einmal in dieser und das andremal in jener Weise aufgefaßt*)."[13] Two *objects*, a Spanish woman and the property of being Spanish, determine the identity of the state of affairs

11 In Husserl *LU* II/1 (V, § 36) 476, (V, § 40) 489, 492, Husserl calls the sense of a declarative sentence 'judgement'.

12 Cp. Husserl *LU* II/1 (IV, § 11) 323-324, (V, § 33) 459-460; Künne 2003, 7–12, 141–145, 252–253, 257.

13 Husserl 1984, 287. In judging that a>b one conceives of the same state of affairs as in judging that b<a, but one does so in a different manner: Husserl *LU* II/1 (I, § 12) 48).

that is (according to Husserl) the intentional object of both John's and Carmen's mental acts, but since at least one of these objects is given to John and Carmen in very different ways, the contents of their acts differ.[14]

Sometimes Husserl uses '*Sachverhalt*' and '*Tatsache*' as stylistic variants of each other,[15] and in ordinary language these two terms are often interchangeable: if you are asking for the explanation of a certain state of affairs then it is a fact that you want to be explained.[16] My impression is that for non-philosophers there is no such thing as a non-obtaining state of affairs: they think that for a state of affairs to exist is to be a fact. Meinong concurs: "that the diagonals in a square are unequal and that sharks are mammals can hardly be called states of affairs (*dass die Diagonalen im Quadrat ungleich oder die Haifische Säugetiere seien, wird man doch kaum Sachverhalte nennen können*)."[17] Be that as it may, a *fact* cannot be the intentional object of a propositional act as a whole unless the content of that act is *true*. Now suppose

(D) Diego judges that Voltaire was dull.

Does his mistaken judgement as a whole have an intentional object? Meinong's answer is Yes. For every judgement and for every non-committal counterpart of a judgement (for every *Annahme*) there is an *Objektiv* at which it is directed, and in the case of Diego's judgement it is an *untatsächliches Objektiv*, a non-factual state of affairs.[18] I think Husserl would have done well to deny this. Just as some

---

14 If Carmen thinks that François Marie Arouet was witty her judgment has a different content than Evita's judgement, but since Monsieur Arouet is identical with Voltaire both acts are correlated with the same state of affairs. Two *objects*, a French philosopher and a property, determine the identity of the state of affairs that is (according to Husserl) the intentional object of both Evita's and Carmen's acts, whereas it is *concepts* under which those objects fall that determine the identity of the propositions that are the contents of their acts.
15 See e.g. Husserl *LU* II/1 (V, § 33) 461.
16 Sometimes Husserl speaks of propositional acts as intending a *Gegenständlichkeit* (objectuality) rather than a *Gegenstand im engeren Sinne* (object in the narrower sence): cp. Husserl *LU* II/1 (I, § 9) 38 n. By contrast, I use 'object' in such a way that it is true that $\forall x$ (x is an object). Sometimes Husserl reserves the term '*Tatsache*' for those states of affairs in which the objects that are thus and so are potential objects of perception: cp. Husserl, 1979, 259–260. By contrast, I employ 'fact' in such a manner that the statements 'The fact that the sun is larger than the moon is identical with the fact that the moon is smaller than the sun' and 'The fact that $2^4<18$ is identical with fact that $18>2^4$' are *both* true. As for Husserl's use of '*Tatsache*' in *LU* see especially vol. II/1 (V, § 33) 460–461 and (V, § 36) 471–476.
17 Meinong 1910, 101.
18 Cp. Meinong 1910, ch. 3 (esp. 95, 101), 240, 277; Meinong 1915, 39, 91. The *Objektiv* that p is non-factual just in case the *Objektiv* that not-p is factual.

but by no means all acts of *non*-propositional thinking have an intentional object, so some but by no means all acts of *propositional* thinking have an objective correlate. Under the (dubious) assumption that propositional acts have their own objective correlates, this correlate can only be something that is the case, it can only be a state of affairs in the colloquial sense of this phrase, that is, it can only be a fact. Varying his formulation in passage (Huss-2) Edmund could and should say: In cases like (D) the judgmental intention exists but *not* the 'intended' state of affairs, whereas in cases like (E) both the judgmental intention *and* the 'intended' state of affairs exist. Since Husserl does not maintain this, his anti-Meinongianism is only half-hearted. So what I suggest at this point is *russelling* Husserl a bit. In his London lectures on Logical Atomism, Russell remarked: 'it does not seem to me very plausible to say that in addition to facts there are also these curious shadowy things going about such as "That today is Wednesday" when in fact it is Tuesday.'[19] Actually my claim is more restricted: no matter whether there are such curious shadowy things or not, no matter whether they deserve such derisive epithets, the assumption that there are non-obtaining states of affairs is no more needed for characterising the structure of *propositional* intentional acts than the assumption that there are Olympian Gods and 1000-faced regular polygons is needed for characterising the structure of *non*-propositional intentional acts.

Obviously, for the specification of the truth-conditions of sentences like 'N. N. thinks that *p*' it is irrelevant whether it is a fact that *p*. So we can specify the truth-conditions of sentences like (E) and (D) along the lines of my reconstruction of Husserl's account of ascriptions of *non*-propositional acts and states. Once again I use square brackets as a singular-term forming operator that transforms an expression into a designator of the sense of that expression. This time the sense is a proposition:

(ε) $\exists x$(x is an act of judging & Evita is the agent of x & the content of x is [Voltaire was witty])

(δ) $\exists x$(x is an act of judging & Diego is the agent of x & the content of x is [Voltaire was dull]).

If somebody judges or believes that things are thus and so then she acknowledges something as true, and this something is the content of her act or state, a proposition, for it is propositions, not states of affairs, that are claimants to the title 'true'. By definition of the square-brackets operator, the result of enframing

19 Russell 1956, 223.

a sentence is free from the state-of-affairs/proposition ambiguity that plagues that-clauses. In asserting (ε) or (δ) we do not ascribe to Evita or Diego a relation to a state of affairs but rather a relation to a proposition: we say of a certain proposition that is the content of somebody's thinking. But if Husserl is right in invoking the category of states of affairs, then in the case of (ε) a relation between Evita and a state of affairs, sc. a fact, does obtain, but in asserting (ε) we do not *state* that this relation obtains.

Why is Husserl's anti-Meinongianism only half-hearted? I shall conclude with some ruminations about this question. Only nine days after he had received a copy of Meinong's book *Über Annahmen* ([1]1902) Husserl sent a rather ungrateful letter to Graz. He claimed that what Meinong calls an *Objektiv* is nothing but what he had called a *Sachverhalt*, he finds it important to mention that in the 1901 edition of *LI* he occasionally called a state of affairs "*das Objektive des urteilenden Vermeinens*", and he goes on to accuse Meinong of plagiarizing.[20] The accusation is demonstrably baseless, but the identification is correct. Furthermore, in their use of the term *Sachverhalt* alias *Objektiv* Meinong and the author of the *LI* both suffer from the same confusion. This confusion was recognized by Husserl's most gifted student in Göttingen. Adolf Reinach severely criticizes Meinong because of his "running together propositions and states of affairs. States of affairs obtain or fail to obtain, propositions are true or false."[21] This is a well-taken criticism, for Meinong does indeed conceive *Objektive* as something the obtaining of which confers truth on certain truth-candidates, and as something that has (if grasped in an act of judging or assuming) a truth-value.[22] But it is hardly possible for one entity to play both roles. "All Austrians", the Rhinelander Reinach complained, "constantly confuse proposition and state of affairs."[23] My teacher Gadamer recalled that in Husserl's seminars and lectures in Freiburg the professor's *Austrian* accent was still unmistakeable and very charming. As Reinach noticed, in the first edition of *LI* the Austrian Husserl repeatedly speaks of *Sachverhalte* as something that is true or false (*etwas, das wahr oder falsch ist, das gilt oder nicht gilt*),[24] and alas! Husserl did not manage

---

20 Husserl (3) 142. The quotation is from Husserl, *LU* II/1 (V, § 28) 445. Husserl might as well have quoted Husserl *LU* II/1 (II, § 23) 164. As for „nine days", see Schuhmann 1977, 71.

21 Reinach 1989 (1911) 114 n., Reinach 1982, 374 n. 21.

22 Cp. Meinong 1910, ch. 3, esp. 94, 100.

23 Cp. Künne 1987, 185–186 for the reference and for a weighting of the advantages, and the disadvantages, of Reinach's canonical names of states of affairs. (Reinach takes it that the use of terms like 'the being witty of Voltaire' (or rather 'Voltaire's being witty') can serve to shield off the "Austrian confusion".)

24 Reinach 1989, 85.

to delete all symptoms of the 'Austrian' confusion in the second edition of the book.[25] Now there are true *and* false propositions, and for each propositional act or state there is a proposition that is its content. If one is not on one's guard against the confusion whipped by Reinach a projection is tempting: There are obtaining and non-obtaining states of affairs, and for each propositional act or state there is a state of affairs that is its intentional object.

At exactly one point in the second edition *LI* Husserl speaks of *wahrmachende Sachverhalte*, truth-conferring states of affairs.[26] Obviously a *Sachverhalt* to which a truth-candidate owes its truth cannot itself be a truth-value bearer, and in order to make something true such a *Sachverhalt* has got to be a state of affairs that obtains, that is, a *fact*. And surely there is no need to postulate non-obtaining states of affairs for the job of making some truth-candidates false. The falsity of (the content of) Diego's judgement about Voltaire is due to a fact, namely to the fact that Voltaire was *not* dull. So whatever the postulation of non-obtaining states of affairs may be good for, they are neither required for characterizing the structure of propositional mental acts and states, nor are they needed for explaining what the falsity of the contents of some propositional acts or states consists in.

As it so happens I think that when in comes to explaining what truth consists in *facts* are not required either. Instead of saying that certain facts make the content of Evita's judgement true and that of Diego's judgement false we can simply say, following Bolzano. 'What she thinks is true *because* Voltaire was witty, and what he thinks is false *because* Voltaire was not dull. Nothing is lost here if we refrain from invoking facts. But that is a long story I told on another occasion.

4 References

Anscombe, E., 1981, 'The Intentionality of Sensation' (1965), repr. in her *Collected Papers*, vol. 2, Oxford, Basil Blackwell, 3-20.

Arnaud, R. B., 1975, 'Brentanist Relations', in: K. Lehrer (ed.), *Analysis and Metaphysics*, Dordrecht, Reidel, 189-208.

Husserl, E., 1979, *Aufsätze und Rezensionen* (1890-1910), ed. B. Rang, The Hague.

Husserl, E., 1980, *Logische Untersuchungen* (^{2}1913/1921), repr. Tübingen.

25 Husserl *LU* II/1 (I, § 11) 43 f, (V, § 10) 367, (V, § 20) 412, (V, § 31) 454.

26 Husserl *LU* II/2 (VI, § 39) 124.

Husserl, E., 1984, Einleitung in die Logik und Erkenntnisthorie. Vorlesungen 1906/07, ed. U. Melle, The Hague.

Husserl, E., 1990-91, 'Intentionale Gegenstände' (1894), ed. K. Schuhmann, in: *Brentano Studies* 3, 137-176.

Husserl, E., 1994, Letter to Alexius v. Meinong (05. 04. 1902), in: *Husserl Briefwechsel*, Dordrecht, vol. 1, 139-145.

Husserl, E., 2011, *Logical Investigations* (LI), transl. by John Niemeyer Findlay (1970), ed. Dermot Moran, London.

Künne, W., 1983, *Abstrakte Gegenstände. Semantik und Ontologie*, Frankfurt/M, (2nd enlarged edn. 2007).

Künne, W., 1986, 'Husserl: Intentionalität', in: J. Speck (ed.), *Grundprobleme der großen Philosophen. Philosophie der Neuzeit IV*, Göttingen 1986, 165-215.

Künne, W., 1987, 'The Difference Between States of Affairs and Propositional Matter', in: Mulligan, 1987, 175-187.

Künne, W., 1991, 'The Nature of Acts: Moore on Husserl', in: D. Bell & N. Cooper (ed.), *The Analytic Tradition*, Oxford, 104-116.

Künne, W., 1995, 'Fiktion ohne fiktive Gegenstände', repr. in: M. E. Reicher (Hg.), *Fiktion, Wahrheit, Wirklichkeit*, Paderborn, 2007, 54-72.

Künne, W., 2003, *Conceptions of Truth*, Oxford.

Künne, W., 2011, '"Denken ist immer Etwas Denken." Bolzano und der (frühe) Husserl über Intentionalität', in: Chr. Beyer & K. Cramer (ed.), Edmund Husserl 1859-2009, Berlin-New York, 77-99.

Meinong, A., 1904, 'Über Gegenstandstheorie', in: Meinong, 1968-78, vol. 2, 481-536.

Meinong, A., 21910, *Über Annahmen*, in Meinong, 1968-78, vol. 4.

Meinong, A., 1915, *Über Möglichkeit und Wahrscheinlichkeit*, in Meinong, 1968-78, vol. 6.

Meinong, A., 1968-78, *Meinong Gesamtausgabe*, ed. R. Chisholm & R. Haller, Graz.

Mulligan, K. (ed.), 1987, *Speech Act and Sachverhalt. Reinach and the Foundations of Realist Phenomenology*, Dordrecht.

Reinach, A., 1989, 'Zur Theorie des negativen Urteils' (1911), in: *Sämtliche Werke*, München, 95-140. Translated as Reinach, Adolf, 1982, 'On the Theory of Negative Judgements', in: B. Smith (ed.) *Parts and Moments*, München, 315-377.

Russell, B., 1956, 'The Philosophy of Logical Atomism' (1918), in his *Logic and Knowledge*, London, 177-281.

Schuhmann, K., 1977, *Husserl-Chronik*, Den Haag.

Twardowski, K., 1982, *Zur Lehre vom Inhalt und Gegenstand der Vorstellungen*, Wien 1894; repr. München.

Christinan BERMES

Commitment (*Verbindlichkeit*)

Intentionality as a Form of Description of the Human World

1 Introduction: Back to the Effectiveness of Intentionality

In the following, I will try to remove the question of intentionality from its self-evident contexts to ask it anew and in a slightly different way. Intentionality shall be defined as an act of committing to something (*sich festlegen*), which draws our attention to the notion of commitment and leads us to culture as the human world. Nowadays, it seems to me, the question of intentionality is not an easy one to ask. Against the background of the extensive and detailed discussions in the fields of philosophy of mind, action theory, epistemology, and social philosophy, numerous problems blend with each another and have led to a rather ambiguous concept of intentionality. We are concerned with various closely linked questions at this point: Is intentionality a property of mind? And if so, what does "property" mean here and how do we have to understand "mind" in this context? Is there a broader definition of intentionality than "consciousness of something" and a narrower term than "intention"? Is it not possible to consolidate both terms in the sense that the intentional and directed consciousness is intention after all? Furthermore, in which medium can intentionality and intentions be realized? Is language the only medium that serves this purpose or are they also expressed in our consciousness, the body (*Leib*), and our actions? Towards what are intentions directed, what are their objects? Are they directed towards particular things or issues, respectively, or do they refer to ideas and representations? Moreover, in what sense can we ascribe existence to intentional objects? Are they existent, non-existent, or existent in their non-existence? Can we describe intentions only from a first-person perspective or can we also describe them from a third-person perspective without a loss of meaning? Are causes of human behavior or purposes of our actions enunciated in intentions?

This list of questions, which is far from being complete, could easily be extended and expressed in a much more detailed and concrete form. Several significant debates of 20th century philosophy, neither of which can solely be ascribed to language-analytic philosophy or phenomenology, are closely

https://doi.org/10.1515/9783110560282-057

connected to these questions. One could even say that 20th century philosophy is, to a large extent, a struggle for intentionality, although the concept itself is not always mentioned explicitly. Certainly, authors such as Husserl, Searle, or Anscombe make direct use of the term, but the problem of intentionality, it seems, is not restricted to the usage of the term. After all, the discussions between Husserl and Frege but also between Frege, Russell, and Wittgenstein concerning the adequately posed question of meaning equally reveal a strong interest in the question of intentionality. The subsequent search for the correct form of a theory of meaning can accordingly also be understood as the search for an adequate theory of intentionality. Every correspondence theory of meaning and every critical assessment of these theories, we could argue, simultaneously comprise a theory of mind and a theory of intentionality. If we had to identify a general goal of the philosophy of the twentieth century – and Michael Dummett is certainly right at this point[1] - the keyword meaning would most certainly come to mind and the task of developing a theory of meaning. What is often forgotten, however, is that there is always a theory of intentionality intrinsically connected to this.

My introductory comments thus lead to the preliminary conclusion that, on the one hand, the discussions about intentionality (in the phenomenological as well as the analytical tradition) are differentiated to such an extent that we run the risk of losing intentionality as a phenomenon. On the other hand, intentionality is at stake in places where it is not explicitly foregrounded as a term – in theories of meaning, for instance. Finally, we put too much confidence in the knowledge we assume to have of the infrastructure that intentionality is grounded on or of the objects that are characterized by intentionality. The mind (*das Geistige*) and consciousness are the most popular candidates to be qualified by intentionality, and with its help, we hope to distinguish between the nature of the mind and non-mental activities (*das Nicht-Geistige*). As Heidegger pointed out in a lecture course in 1925, it was this kind of thinking that trapped the concept of intentionality in a fixed interpretative pattern, at least from a phenomenological perspective. He remarks pointedly:

> In the popular philosophical literature, phenomenology tends to be characterized in the following manner: Husserl took over the concept of intentionality from Brentano; as is well known, intentionality goes back to Scholasticism; it is notoriously obscure, metaphysical, and dogmatic. Consequently, the concept of intentionality is scientifically useless and phenomenology, which employs it, is fraught with metaphysical presuppositions and therefore not at all based upon immediate data.

1 Bermes 1997, 83 ff.

> *(In der philosophischen Tagesliteratur pflegt die Phänomenologie so charakterisiert zu werden: Husserl hat von Brentano den Begriff der Intentionalität übernommen; Intentionalität geht bekanntlich auf die Scholastik zurück; diese ist finster, metaphysisch, dogmatisch. Folglich ist der Begriff der Intentionalität wissenschaftlich unbrauchbar und die Phänomenologie, die ihn verwendet, mit metaphysischen Voraussetzungen belastet, ist also gar nicht auf unmittelbare Gegebenheiten gegründet.)*[2].

Nowadays, Heidegger himself does not necessarily get a good press in "the popular philosophical literature" – with good reason. I will also have to consider the fact that, after the publication of the *Black Notebooks*, the judgement of the author of *Being and Time* must be doubted[3]. But it would also be trite to ban Heidegger from philosophy. After all, the kind of thinking he unfolded in the public spaces of his university lectures seems to be less contaminated than the empty tirades resulting from his soliloquies. Sometimes you even get the impression that his audience had an almost therapeutic effect on Heidegger and saved him from further excesses, if not necessarily at all times.

Regardless of his contemporary standing and his ever-changing importance, Heidegger made some valuable contributions to the concept of intentionality in his lectures in the summer term of 1925, published under the title *History of the Concept of Time: Prolegomena,* which are still relevant for contemporary discussions on the subject. He issues a timeless warning, because Heidegger emphasizes that the "title" of intentionality "is the very last word to be used as a phenomenological slogan" and that intentionality rather "identifies that whose disclosure would allow phenomenology to find itself in its possibilities"[4].

This hint is more than an empty phrase; on the contrary, it affects Heidegger's philosophy as well as Husserl's. For Husserl, intentionality is not a problem that can be solved easily in the sense that you might discover a formula at the end, as in the case of *noese* and *noema*, for example, which necessarily correlate. The concept of intentionality as we find it in Husserl's writings confronts us with a form of description that can be formulated in various ways and structures human experience accordingly. In Husserl's understanding, intentionality does not provide a solution; it rather sets up a task. This also explains why Husserl repeatedly returned to the subject of intentionality throughout his

---

2 Heidegger GA 20, 35. In the following, citations from Heidegger refer to the *Gesamtausgabe* (GA), from Husserl to Husserliana (Hua), and from Wittgenstein to the *Werkausgabe* (the respective work will be stated in the text).

3 Bermes 2016.

4 Heidegger GA 20, 62.

life. Therefore, we should call intentionality a phenomenological project that is meant to reveal something which is not visible from a non-phenomenological perspective.

Considering this, it seems less convincing to use intentionality as a criterion to differentiate between Husserl and his successors, as it is often done in various approaches. One thing that immediately comes to my mind is our understanding of the line of development from Husserl to Heidegger and eventually to Merleau-Ponty and Sartre. Nevertheless, it is not completely wrong to distinguish between several concepts of intentionality to establish a certain order. The picture presents itself as follows: 1. From the very beginning of his phenomenological work, Husserl assumes a universal "a priori of correlation" (*Korrelationsapriori*) which he describes as a unity of intentional consciousness and intentional object. 2. His successors, Merleau-Ponty for instance, follow Heidegger insofar, as they consider intentionality as a conception of world and simultaneously realize that a surplus of meaning is revealed beyond intentionality which cannot be grasped by intentionality. 3. Recent phenomenological approaches, then, are concerned with emerging phenomena that, somehow conversely to intentionality, impose themselves on consciousness and maybe even question it[5].

I would like to stress again that such classifications can be helpful; in this case, however, intentionality is a title or slogan that is used to realize particular theories but not to access the phenomena themselves. To return to the essence of intentionality, it is thus necessary to consider it in its own effectiveness. There are certainly things we may find problematic with regard to the structure of intentionality, but it cannot be denied that it is part of our life and affects it as well.

This being said, my approach should be clear, and you could further illustrate this with a well-known remark by Wittgenstein:

> We have got on slippery ice where there is no friction and so in a certain sense the conditions are ideal, but also, just because of that, we are unable to walk. We want to walk: so we need friction. Back to rough ground!
> (*Wir sind aufs Glatteis geraten, wo die Reibung fehlt, also die Bedingungen in gewissem Sinne ideal sind, aber wir eben deshalb auch nicht gehen können. Wir wollen gehen; dann brauchen wir die Reibung. Zurück auf den rauhen Boden!*)[6].

---

5 Such a typology can be found in Gondek, Tengelyi 2011, 664-675.

6 Wittgenstein PU, § 107.

To get intentionality back to rough ground and to use the resulting friction for our understanding of intentionality, we need to make its effectiveness explicit. In doing so, we can discover a simple fact: Whoever desires something is also committed to something (*sich festlegen*). People who aim at something, perceive something, or buy something are committed to something at the same time. Their commitment binds them to the objects they desire (or aim at, perceive...) and the act of desiring something. Commitments are related to the act of taking a stance. If you are committed to something, you also take a stance towards something (*Stellung beziehen*), you prepare yourself for something, you respond to something. My first claim is that intentionality is the form of accessing the world which enables us to understand ourselves and the world through the acts of taking a stance on and committing to something. Committing to something is not a form of blind insistence or rigid fixation, it rather means that we get involved with an intentional practice that contains and shows something in itself. Accordingly, intentionality can be grasped as a title that deciphers the human world – it is human beings who understand themselves through their acts of committing to something and taking a stance. If we understand intentionality as a form of commitment, a particular aspect of Husserl's phrase "consciousness as consciousness of something" arises. This phrase implies, in my reading, that that of consciousness commits to itself. I will use the term commitment (*Verbindlichkeit*) to describe the complex structures we find among the diverse acts of committing and taking a stance.[7] This brings me to another thesis, which is that a theory of intentionality can be extended to a theory of commitment. The terms "taking a stance", "committing to something", and "commitment" additionally have the advantage that they are not restricted to either the sphere of the practical or the theoretical but encompass both and can serve integrating purposes.

To understand the theory of intentionality as a theory of commitment, I will firstly elaborate on and interpret some (by no means all) aspects of intentionality and then continue my analysis with a discussion of Husserl and Wittgenstein. What I would like to suggest is that Husserl's reflections on life world (*Lebenswelt*) as well as Wittgenstein's on certainty express a common interest: They are both interested in the commitment of our natural understanding of the world. Finally, I will consider the consequences of these thoughts by assigning a new place to the question intentionality. This question may be asked correctly in

7 The German term "*Verbindlichkeit*" designates the fact that we are bound to the object of our commitment. It could be translated as well as "engagement", in the sense of an attitude we adopt when we are dedicated to a particular activity.

philosophy of mind, it certainly has earned its rightful place in epistemology, and it can be located at the boundaries of metaphysics, but cultural philosophy might be the only discipline that can provide an answer to it.

My argument is tentative and has an experimental trajectory. I see Husserl, Wittgenstein, and Heidegger as interlocutors with different opinions on many issues, but who agree on the fact that intentionality must be understood in its effectiveness and that it does not promise any solutions but instead expands our horizon. To say that my trajectory of my argument is experimental does of course not mean that I do not take a serious stance; experiments are serious things after all. But I will not offer a Wittgenstein or Husserl philology, and some aspects may be left unsaid although you actually had to mention them.

2 Intentionality, Determination, and Positioning

In his *Philosophical Investigations* but of course also in the *Philosophy of Psychology* and many other works, Wittgenstein struggles to uncover something very simple and self-evident – our action (*das Tun*):

> I should not say of the movement of my arm, for example, it comes when it comes, etc. And this is the region in which we say significantly that a thing doesn't simply happen to us, but that we do it.
> (*Von der Bewegung meines Armes, z.B., würde ich nicht sagen, sie komme, wenn sie komme etc. Und hier ist das Gebiet, in welchem wir sinnvoll sagen, dass uns etwas nicht einfach geschieht, sondern dass wir es tun*)[8].

We might be moved from one room to another in a hospital, spin through the air in an amusement park, or be worried about a strong heartbeat. In all these cases we can be surprised[9] and wonder what exactly is or has been happening to us. However, if we extend this pattern of questions and explanations to the relationship between will and action, we are confronted with a serious problem. If somebody writes an email to his secretary, he cannot answer the question about what he is doing by stating that something is happening to his hands, meaning that his will affects his hands (as the nurse affected the bed in the hospital at night by moving it to another room). In this particular case everything is absent

8 Wittgenstein PU, § 612.
9 "So one might say: voluntary movement is marked by the absence of surprise" ("Man könnte also sagen: die willkürliche Bewegung sei durch die Abwesenheit des Staunens charakterisiert") Wittgenstein PU, § 628.

that we usually associate with actions that happen to us, for instance a feeling of surprise about what is happening. We also lack a clear connection to the instruments which possibly provide an explanation for what is happening but do not clarify the action:

> When I raise my arm 'voluntarily', I do not use any instrument to bring the movement about.
> (Wenn ich meinen Arm ‚willkürlich' bewege, so bediene ich mich nicht eines Mittels, die Bewegung herbeizuführen)[10].

In the writing process nothing is happening to us, instead the writer determines himself in the act of writing by writing something and insofar as he writes something.

Let us consider this example a little closer. If somebody asks us what it means to write an email to the secretary, we could answer:

> He writes an email to his secretary by competently using the computer and entering a text.

But we could also say:

> He writes an email to his secretary insofar as he competently uses the computer and enters a text.

Undoubtedly, there is an intentional issue at stake here. The writing process as *intentio* is directed towards the email to the secretary as its *intentum*, its intentional object. This intentional object is an intentional object precisely because it stands in an appropriate relation to the act itself – i.e., the act of writing. It is also insignificant if the email is sent or not, this does not affect the action. It is insignificant whether or not the email reaches the secretary if we want to understand the action we are interested in at this point. The same is true for the act of shooting. To understand the act of shooting it is not necessary that the stag is actually hit. It can still be called the act of shooting at the stag even if we miss the stag[11]. The logic of shooting allows us to miss.

If we wish to explain the act of writing an email to the secretary we either indicate the manner of the act or how the act is done (by competently using the computer and entering a text), or the condition of the action (insofar as he com-

10 Wittgenstein PU, § 614,

11 Anscombe uses the example of shooting at the stag to define the intentional object in her understanding. See Anscombe 2014, 230–260.

petently uses the computer and enters a text). I intend to return briefly to both terms at a later point.

Let me elaborate this case a little further: If somebody watches this situation and sees that the address given in the header field is not that of the secretary but of a friend, the spectator could say: Actually, the email is not addressed to the secretary, it is addressed to a friend. If this is a valid description of the situation, the expression "to a friend" does not denote the intentional object, because "the friend" does not stand in the appropriate relation to the act of "writing an email to the secretary". In scholasticism you could call this the "material object" as distinct from the intentional object of "the email which is addressed to the secretary". In phenomenology, "the friend" would not function as the intentional object either because according to phenomenology, we had to admit that "the friend" cannot fulfill the intention. The intention, after all, is qualified by the intentional object of "the email which is addressed to the secretary".

There are of course several problems connected to the description and analysis of this situation which I will not mention here. Instead, I will continue with phenomenology, since this language-analytical consideration immediately leads to phenomenology and grasps intentionality. As Heidegger remarks on the intentionality of perception:

> The perceived in the strict sense for phenomenology is not the perceived entity in itself but the perceived entity insofar as it is perceived, as it shows itself in concrete perception
> (*Das Wahrgenommene im strengen Sinne der Phänomenologie ist nicht das wahrgenommene Seiende an ihm selbst sondern das wahrgenommene Seiende, sofern es wahrgenommen ist, wie es sich in konkreter Wahrnehmung zeigt*)[12].

This quote elucidates Husserl's formula of intentionality as "*cogito-cogitatum* (*qua cogitatum*)"[13]:

> Its beginning is the pure – and, so to speak, still dumb – psychological experience, which now must be made to utter its own sense with no adulteration. The truly first utterance, however, is the Cartesian utterance of the ego cogito – for example: I perceive – this house or I remember – a certain commotion in the street. And the first descriptive generality is the separation of *cogito* and *cogitatum* qua cogitatum.
> (*Der Anfang ist die reine und sozusagen noch stumme Erfahrung, die nun erst zur reinen Aussprache ihres eigenen Sinnes zu bringen ist. Die wirklich erste Aussprache ist aber die Cartesianische des ego cogito, z.B.: Ich nehme war – dieses Haus wahr, ich erinnere mich –*

12 Heidegger GA 20, 52.
13 Husserl Hua I, 74.

> *eines gewissen Straßenauflaufs usw.; und das erste Allgemeine der Beschreibung ist die Scheidung zwischen cogito und cogitatum qua cogitatum*)[14].

An almost trivial fact that has commonly been overlooked in the reception of phenomenology is that this structure of intentionality is not a property of an action or an experience which is attributed to this action or experience from the outside. Also, the intention is not carried to the object, so that you could firstly experience the object and then, in a second step, the intentionally constituted object:

> There are not two things present in experience, we do not experience the object and beside it the intentional experience directed upon it ... only one thing is present, the intentional experience, whose essential descriptive character is the intention in question.
> (*Es sind ... nicht zwei Sachen erlebnismäßig präsent, es ist nicht der Gegenstand erlebt und daneben das intentionale Erlebnis, ... sondern nur Eines ist präsent, das intentionale Erlebnis, dessen wesentlicher Charakter eben die bezügliche Intention ist*)[15].

Heidegger's formulation and understanding of the *cogitio-cogitatum* (*qua cogitatum*) is very precise if it comes to the intentional object. For him, the intentional object is likewise qualified by modality (how it shows itself in concrete perception) and condition (if it is perceived). With regard to our example, this means that it is not the email as an incident among others and of course also not the representation of the email (because we do not write representations) which serve as the intentional object of the act of writing, but the email to the secretary how or as it shows itself in the concrete act of writing and insofar as it is written. This, however, precisely marks the description of the act of writing as a form of commitment. The act of writing is informed by a commitment to the process of writing and its object is defined by this commitment.

In his *Philosophical Grammar*, Wittgenstein describes this circumstance as follows:

> A search for a particular thing (e.g. my stick) is a particular kind of search, and differs from a search for something else because of what one does (say, thinks) while searching, not because of what one finds.
> (*Ihn (etwa meinen Stock) suchen, ist eine Art des Suchens und unterscheidet sich davon, dass man etwas andres sucht, durch das, was man beim Suchen tut (sagt, denkt), nicht durch das was man findet*)[16].

---

14 Husserl Hua I, 77.
15 Husserl Hua XIX/1, 386.
16 Wittgenstein PG, § 91.

He continues:

> The idea that it takes finding to show what we were looking for, and fulfilment of a wish to show what we wanted, means one is judging the process like the symptoms of expectation or search in someone else.
> (*Der Gedanke, dass uns erst das Finden zeigt, was wir gesucht, erst die Erfüllung des Wunsches, was wir gewünscht haben, heißt den Vorgang so beurteilen, wie die Symptome der Erwartung oder des Suchens bei einem Andern.*)[17]

However:

> The symptoms of expectation are not the expression of expectation.
> (*Die Symptome der Erwartung sind nicht der Ausdruck der Erwartung.*)[18]

What Wittgenstein says here can also be said of intention: It "isn't something I see happen" but "it's more like my being involved in my actions, my being my actions"[19].

Following this, we could say that intentionality consists of a particular kind of commitment within a practice (*Sich-in-Praxis-Festlegen*), how something is done and insofar as it is done – at this point precisely it is "my being involved in my action". By doing so we determine ourselves in our actions: According to Wittgenstein, we follow certain rules, and to speak with Husserl, we are engaged in our intentional life of action (*Aktleben*). As Merleau-Ponty stresses, the intentional consciousness is "originally not a 'I think to...' but a 'I can'"[20] (ursprünglich nicht ein 'Ich denke zu...' sondern ein 'Ich kann').

3 Intentionality, Commitment, and Culture

At the beginning of my paper I said that a theory of intentionality, as a form of committing to something, can be developed into a theory of commitment, and I suggested that we can illustrate this by looking at Wittgenstein and Husserl. Due to a lack of time and space, I can only sketch one possible way and give some hints about how this could be achieved. The rationale behind this is comparatively simple one: Intentionality can be seen as the act of committing to

17 Wittgenstein Zettel, § 57.
18 Wittgenstein PG, § 92.
19 Wittgenstein PG, § 97.
20 Merleau-Ponty 1966 (1945), 166.

something in an action by realizing the action and engaging in a particular manner. In turn, these actions refer to habits, customs, and institutions. We could also say that they are embedded in customs and institutions:

> An intention is embedded in its situation, in human customs and institutions. If the technique of the game of chess did not exist, I could not intend to play a game of chess. In so far as I do intend the construction of a sentence in advance, that is made possible by the fact that I can speak the language in question.
> (*Die Absicht ist eingebettet in die Situation, den menschlichen Gepflogenheiten und Institutionen. Gäbe es nicht die Technik des Schachspiels, so könnte ich nicht beabsichtigen, eine Schachpartie zu spielen. Soweit ich die Satzform im voraus beabsichtige, ist dies dadurch ermöglicht, dass ich deutsch sprechen kann*)[21].

In his *Blue Book*, Wittgenstein develops an interesting metaphor for institutions. He describes the use of the word 'meaning' as a 'worker doing odd jobs' within institutionalized labor. Occasionally, therefore, we need to ask what is meant or what could be meant. If language works, meaning or interpreting do not interfere:

> Meaning is one of the words which one may say that they have odd jobs in our language. It is these words which cause most philosophical troubles. Imagine some institution: most of its members have certain regular functions, functions which can easily be described, say, in the statutes of the institution. There are, on the other hand, some members who are employed for odd jobs, which nevertheless may be extremely important. What causes most trouble in philosophy is that we are tempted to describe the use of important 'odd-job'-words as though they were words with regular functions.
> (*‚Meinen' ist eines der Wörter, von denen man sagen kann, dass sie Gelegenheitsarbeiten in unserer Sprache ausführen. Diese Wörter verursachen die meisten philosophischen Verwirrungen. Stell dir irgendeine Institution vor: die meisten Mitglieder haben bestimmte, regelmäßige Funktionen, - Funktionen, die etwa in den Statuten der Institutionen leicht beschrieben werden können. Andrerseits gibt es einige Mitglieder, die für Gelegenheitsarbeiten, die jedoch äußerst wichtig sein können, angestellt werden. Unsere Neigung, den Gebrauch wichtiger ‚Gelegenheitsarbeiter'-Wörter so zu beschreiben, als ob sie Wörter mit regelmäßigen Funktionen wären, stiftet die meiste Verwirrung in der Philosophie*)[22].

The metaphor of the 'odd-job worker' (*Gelegenheitsarbeiter*) who is employed in an institution in which other workers do regular jobs allows us to speak of 'commitment'. Institutions create commitment by providing regular jobs, while we commonly look for commitment at the wrong place, i.e. at odd-job workers. When Husserl addresses normality and abnormality, he discusses this issue:

---

21 Wittgenstein PU, § 337.
22 Wittgenstein BB, p. 74.

> It is reality that underlies everything, the descriptive and actual world of a system of normal experiences which are characterized in themselves as immediately giving reality; and this is what abnormal experiences refer to.
> (*Zugrunde also liegt die Wirklichkeit, die anschauliche und wirkliche Welt eines Systems normaler Erfahrungen, die in sich selbst als unmittelbar Wirklichkeit gebende charakterisiert sind; und darauf bezogen sind die anomalen Erfahrungen.*)[23]

After my former remarks, we can interpret Wittgenstein's metaphor in the sense that Husserl in the analysis of life world surrounding his *Crisis* and Wittgenstein in *On Certainty* both try to illustrate the architecture of an institution or the function of institutions in general. They do not want to focus on the odd-job workers but on the work that is done on a regular basis, because it is this kind of work that guarantees commitment. Therein, our intentional self- and world-orientation shows a structure marked by 'certainties', 'secure convictions', and 'safeties' in the case of Wittgenstein, and by a net of convictions that document themselves in practice in the case of Husserl. As Husserl states:

> The practice is limited in its particularity, which shall be regarded as an (veritable, theoretical) entity and a true judgement.
> (*Praxis in ihrer Jeweiligkeit begrenzt, was als (wahrhaft, theoretisch) Seiendes und wahres Urteil gelten soll.*)[24]

Wittgenstein, then, writes:

> Giving grounds, however, justifying the evidence, comes to an end; - but the end is not certain propositions' striking us immediately as true, i.e. it is not a kind of seeing on our part; it is our acting, which lies at the bottom of the language-game.
> (*Die Begründung aber, die Rechtfertigung der Evidenz kommt zu einem Ende; – das Ende aber ist nicht, dass uns gewisse Sätze unmittelbar als wahr einleuchten, also eine Art Sehen unsrerseits, sondern unser Handeln, welches am Grunde des Sprachspiels liegt.*)[25]

In their advanced philosophy, Wittgenstein as well as Husserl provides a certain kind of archaeology of culture by identifying certainties as rules of practice at the bottom of our knowledge. Such certainties are far from being subjective. In fact, their reference point is not a subject but a shared practice. As such, they can only be described[26] and give "form" to our claims to knowledge[27]. Husserl

23 Husserl Hua XXXIX, 641.
24 Husserl Hua XXXIX, 200.
25 Wittgenstein ÜG, § 204.
26 Wittgenstein ÜG, § 189.
27 Wittgenstein ÜG, § 211.

does not use the term "form" at this point, instead he speaks of "style" or "typology" that allow for unanimity. Wittgenstein, then uses several metaphors to illustrate the relations between knowledge and certainty:

> I do not explicitly learn the propositions that stand fast for me. I can discover them subsequently like the axis around with a body rotates. This axis is not fixed in the sense that anything holds it fast, but the movement around it determines its immobility.
> (*Die Sätze, die für mich feststehen, lerne ich nicht ausdrücklich. Ich kann sie nachträglich finden wie die Rotationsachsen eines sich drehenden Körpers. Diese Achse steht nicht fest in dem Sinne, dass sie festgehalten wird, aber die Bewegung um sie herum bestimmt sie als unbewegt.*)[28]

These practiced certainties could indeed be called "culture", although Wittgenstein uses the term "world-picture". A world-picture is not a part of our knowledge of whose correctness we have been or could be convinced, it is "the inherited background against which I distinguish between true and false"[29].

> The propositions describing this world-picture might be part of a kind of mythology. And their role is like that of rules of a game; and the game can be learned purely practically, without learning any explicit rules.
> (*Die Sätze, die dieses Weltbild beschreiben, könnte man mythologisch nennen. Und ihre Rolle ist ähnlich der von Spielregeln, und das Spiel kann man auch rein praktisch, ohne ausgesprochene Regeln lernen.*)[30]

This world-picture cannot be given to somebody „through a kind of persuasion"[31], it is neither an "invention" nor a "hypothesis"[32] – it is lived, so to speak. Certainly, the world-picture can change, but not in the sense that one world-picture is exchanged for another. However, individual convictions within a world-picture can change their functions:

> It might be imagined that some propositions, of the form of empirical propositions, were hardened and functioned as channels for such empirical propositions as were not hardened but fluid; and that this relation altered with time, in that fluid propositions hardened, and hard ones became fluid.
> (*Man könnte sich vorstellen, dass gewisse Sätze von der Form der Erfahrungssätze erstarrt wären und als Leitung für die nicht erstarrten, flüssigen Erfahrungssätze funktionierten; und*

28 Wittgenstein ÜG, § 152.
29 Wittgenstein ÜG, § 94.
30 Wittgenstein ÜG, § 95.
31 Wittgenstein ÜG, § 262.
32 Wittgenstein ÜG, § 167.

dass sich dies Verhältnis mit der Zeit änderte, indem flüssige Sätze erstarrten und gefrorene flüssig würden.)[33]

These suggestions should be sufficient to illustrate how, on the basis of Husserl and Wittgenstein, a theory of intentionality can be extended to a theory of commitment (*Verbindlichkeit*). There are of course several differences between both approaches, but you can also recognize a common interest: a common interest in commitment within the interplay of intentional practices.

4 Conclusion

I would like to conclude my paper with the warning issued by Heidegger that the "title" of intentionality should by no means "be used as a phenomenological slogan" since intentionality "identifies that whose disclosure would allow phenomenology to find itself in its possibilities"[34]. To avoid its slogan-like nature, I have suggested that intentionality should be understood as the act of committing on something within an action by realizing itself and engaging with a particular matter. This of course questions the definition we find in the writings of Brentano, Husserl, and others, who thought that the phrase 'being directed towards something' would represent the classical feature to define intentionality. It would be a misunderstanding of my paper, however, if somebody accused me of overlooking the aspect that 'intentions' are always directed towards something. What I am saying is that 'being directed towards something' realizes itself in the form of 'committing on something'. Thus, the distinction between intention and action collapses and this paves the way for discussions about intentionality beyond the realms of philosophy of mind, for instance, in cultural philosophy. It is possible that the outcome of such a concept of intentionality provides new perspectives for cultural philosophy – a cultural philosophy that sees culture as the institution that is created through actions, exists in actions, and brings about commitment in the form of certainty. This culture cannot be "made", although it exists through actions. This concept of culture does not only encompass the acts of committing to something and taking a stance, the form of commitment is of paramount importance for it.

33 Wittgenstein ÜG, § 96.
34 Heidergger GA 20, 62.

5 References

Anscombe, G.E.M., 2014, Die Intentionalität der Wahrnehmung, in: Anscombe, G.E.M., *Aufsätze*, Berlin, Suhrkamp, 230-260.
Bermes, Chr., 1997, *Bedeutung als Bestimmung und Bestimmbarkeit. Eine Studie zu Frege, Husserl, Cassirer und Hönigswald*. Würzburg, Königshausen & Neumann.
Bermes, Chr., 2016, Verrat am Leben? Heidegger und der Tod, in: H.-H. Gander, M. Striet (eds.): *Heideggers Weg in die Moderne, eine Verortung der der "Schwarzen Hefte"*. Frankfurt, Vittorio Klostermann.
Gondek, H.-D., L. Tengelyi, 2011, *Neue Phänomenologie in Frankreich*. Berlin, Suhrkamp.
Heidegger, M., 1910 ff., *Gesamtausgabe* (GA), Frankfurt am Main, Vittorio Klostermann.
Husserl, E., 1950 ff., *Husserliana* (Hua), The Hague, Martinus Nijhoff; Dordrecht, Kluwer Academic Publishers; New York, Springer.
Merleau-Ponty, M., 1966, *Phänomenologie der Wahrnehmung* (frz. 1945), Berlin, de Gruyter.
Wittgenstein, L., 1984 ff., *Werkausgabe*. Frankfurt a.M., Suhrkamp.

Jesús Padilla Gálvez

The Paradox of Intentionality

1 Introduction

What is behind someone's claiming that he or she has understood a person's action? To answer this question, we must first clarify the terminology used in the context of intention, above all the terms meaning and purpose. Since F. Brentano's works the view has deepened that understanding an action is a psychological process which goes on in a person's mind.[1] An example from criminal law shall illustrate the aspects involved in human action. In the investigation of a criminal offence, valid evidence must be based on two conditions: "*actus reus*" and "*mens rea*". According to the legal standards the prosecutor must prove beyond doubt that the suspect ("*actus reus*") and he must also show the suspect's intention of committing the crime ("*mens rea*"). To do this he has to reconstruct the sequence of actions and the manner in which they are linked to intentional conduct.[2] In common law countries the judge must follow the legal principle "*actus non facit reum nisi mens sit rea*" to determine criminal responsibility. This principle means that an action does not make a person guilty unless the mind is also guilty. In criminal trials both, *actus reus* and *mens rea* must be proved before it comes to a conviction. This example shows that the concept of intentionality is used in criminal law to detect and convincingly proof unlawful acts.

One might ask whether action is a process that goes hand in hand with comprehension. In other words, is the process of comprehension related to the enunciation of a proposition? Is it comparable to a process going on in a person's mind when understanding a sentence? Are these questions appropriate to approach the problem or should we rather focus on an agent's underlying intention for an action? We shall change our perspective.

---

1 Cf.: Brentano 1995, 88 f. Brentano 2008, 106.

2 Cf.: Padilla Gálvez 2016, 59 ff.

https://doi.org/10.1515/9783110560282-073

2 Vindicating the Concept of Intentionality

Intention is one of those concepts whose history is well documented. It was first used by the medieval philosopher Ibn Sina (Avicenna) in his book *Kitab al - Isharat wa – 'l – tanbihat* (*Remarks and Admonitions*).[3] Philosophers as well as jurists consequently adopted the concept and used it in new contexts. In order to examine its meaning we will present a thought experiment (*Gedankenexperiment*[4]) that reveals its origin. It is Avicenna's renowned thought example of the 'flying' or 'floating man', who has suddenly come into existence at the age of thirty and floats in space with outstretched limbs but lacks any sense perception. Despite of this incapability the man is still aware of his own existence. The state of floating refers to a situation in which a man is able to think though he has no sense data of a material body. According to Avicenna, the floating man is able to comprehend the concept of being without any sense experience and concludes that humans cannot be denied their consciousness. The conditions of this argument are these: let us suppose a man was just created at a stroke being fully developed and perfectly formed but with a constrained vision of external objects. He is floating in space without noticing any air draft and with his extremities separated so that arms and legs cannot touch. Then let the subject consider whether he would affirm the existence of his self. Despite of being unable to feel his body, there is no doubt that he would affirm his own existence. And if it were possible for him in such a state to imagine a hand or any other organ, he would not imagine it to be a part of himself or a condition of his existence. This argument relies on an introspective thought experiment.[5] In this context Avicenna mentioned the word "ma'nā" (*intentio*) to refer to all those cognitive contents that we know of our vital selves. He described this in the following quote:

> ... quod autem apprehendunt vires ocultae absque sensu, vocatur in hoc loco propriie nomine intentionis.[6]

Intentions entail such evaluative cognitive contents. The term 'intention' goes back to the Arabic word ma'nā[7] and was later translated into Latin as 'intentio'

3 Ibn Sina, 1960 343–346; Ibn Sina 1996.
4 Padilla Gálvez 2010, 113 ff.
5 Ibn Sina 1972; Ibn Sina 1968.
6 Ibn Sina 1972, 86.

by Dominicus Gundissalinus in the Toledo School of Translators.[8] According to Gundissalinus the soul has five different functions, such as "*phantasia*", "*imaginatiua*" (or "*formatiua*"), "*imaginatio*", "*aestimatiua*" and "*memoralis*". The fourth function "*aestimatiua*" is responsible for comprehending intentions ("*appehendens intentiones*") and is located "*in summo mediae concauitatis cerebri*"[9]. He described intention in the following quote:

> Virtus uero apprehendens ab intus duplex est, quoniam alia apprehendit formas sensibilium, alia intentiones sensibilium; interest autem inter apprehendere formam et apprehendere intentionem. [...] Quod uero apprehendunt uires interiores absque sensu uocatur hoc loco intentio.[10]

If one looks at intention in this way it is a skill that is learned internally without any sensory organs. Gundissalinus considered intention as belonging to such skills as estimation and evaluation (*aestimatiua*) and its function is to enhance the ability to learn intentions that cannot be sensually experienced. Almost unnoticed by many scholars this basic definition had a significant influence on Thomas Aquinas' writings:

> esse intentionis intellectae in ipso intelligi consistit.[11]

Thomas Aquinas stated that the *being of an intention consists simply in its being thought* and with this simple definition he placed intentionality outside the sensory experience and linked it to thought. This chapter deals with the ontological thesis of the nature of certain objects of thought. It is based on Avicenna's *Gedankenexperiment* of intentional inexistence, which refers to thoughts that are directed towards inexistent objects. There is a distinction between Avicenna's is thinking about a floating man and Avicenna's thinking about nothing. In the first case Avicenna is supposed to be intentionally related to an object, whereas in the second case he is not. In fact, the status of a floating man cannot correspond to a floating man as men suddenly appearing at the age of thirty and float in space do not exist. Yet the very thought of a floating man as an intentionally inexistent being and as a mental construct lasts longer than the actual thought of it.

---

7 The common source for claims regarding the translation of "maʿnā" and other terms as intention is Gyekye 1971, 32-38.

8 Gundissalinus 1999, 152; 23, 21 f.

9 Gundissalinus 2009, 204; 132-134.

10 Gundissalinus 2009, 202; 131 /71M/. Cf.: Avicenna Latinus 1972, 86, 1.99 ff.

11 Thomae de Aquino 1948, Lib. 4, cap. 11, n. 6.

Intentionality is thus anchored in the argument of the dualism 'body-mind'.[12] Descartes claimed that the mind possesses certain properties which the body does not have and thereby negated a unity of body and mind. The fundament on which this negation is based is Leibniz's principle of the indiscernibility of identical objects. The non-identity of body and mind is linked to the converse intentional property. These are attributes a thinking subject ascribes to objects in accordance with the subject's intentional attitudes. A brief example shall illustrate what was said: if I love Toledo, I have the intentional property of loving Toledo, while Toledo has the inverse intentional property of being loved by me.

In chapter two of Descartes' work we come across a fallacy as the intentional properties described there are all of extrinsic nature. Thus the philosopher ignored the intrinsic property of intention, which had been a characteristic element of intention since Gundissalinus, was ignored. The argument is defective because it is based on an unrestricted and incorrectly generalizing version of the inviolability of identity. The principle of inviolability of identity enables us to determine non-identity on the basis of inverse intentional properties. We are confronted with these kinds of problems when dealing with intentionality.

It was not until three centuries later that Brentano took up these problems again and proposed new solutions. He considered intentionality an essential feature of mental processes and viewed it as characteristic of all acts of consciousness:

> Every mental phenomenon is characterized by what the Scholastics of the Middle Ages called the intentional (or mental) inexistence of an object, and what we might call, though not wholly unambiguously, reference to a content, a distinction upon an object (which is not to be understood here as meaning a thing), or immanent objectivity. Every mental phenomenon includes something as object within itself, although they do not all do so in the same way. In presentation something is presented, in judgement something is affirmed or denied, in love loved, in hate hated, in desire desired and so on.
> This intentional in-existence is characteristic exclusively of mental phenomena. No physical phenomenon exhibits anything like it. We could, therefore, define mental phenomena by saying that they are those phenomena which contain an object intentionally within themselves.[13]

To elucidate the meaning of intention we shall use again an example from criminal law. Suppose the witness of a crime is shown an image of a suspect in the form of a facial composite (*Phantombild*) and while looking at it the witness says

12 Descartes 1964–1974, chapter 2.
13 Brentano 1995, 88 f. Brentano 2008, 106.

"This facial composite represents the criminal".[14] What exactly is the relationship between the image and the actual criminal? Is it the same relationship that exists between the colour particles on the paper and the criminal? One may argue that the facial composite resembles the criminal and this is the link. But what if the facial composite looks deceptively similar to the suspect but the crime was committed by another person? Does the facial composite in that case still refer to the criminal? And what if the facial composite looks deceptively similar to more than one person? In the latter case one would have to create a mental relation between the image and the corresponding persons. Brentano considered processes in which a relation between image and reality is established to take place in consciousness. The mind bestows meaning to the image and assigns symbols to objects.

Brentano's view seems to be based on correct assumptions but we want to ascertain what we do when assigning symbols (*Zeichen*) to objects? According to Brentano several mental operations are involved, among them acts of intention that produce attributing functions. He named the whole process an act of consciousness. The importance of a facial composite is that of attributing, and it is this ability on which our thoughts and thinking rests. The ability of making ascriptions is essential and immediately given and we can only accept it.

How convincing is Brentano's answer? There are some questions that require clarification. For example, the question of *how* an intentional act commences, proceeds and continues. The act of intending is indeed an experience. In our example it is the knowledge of a crime accompanied by a desire to discover the perpetrator. But according to which criteria is this knowledge assigned to reality? How can an event foreshadow another event? The answer from a phenomenological perspective[15] is simple: the peculiarity of mental processes lies in the fact that everyone can only experience them oneself. As some of our questions remain unanswered we shall try another approach to the problem.

Suppose I have heard that the facial composite corresponds to a person called N and I write down N's name in my notebook. What kind of nexus is there

---

14 The Spanish MP Gaspar Llamazares's image was used for computerised mock-ups of Osama bin Laden and another al-Qaida leader. Wittgenstein says disregarding: "But if we wish to know what it means "intending this to be a portrait of so-and-so"." Cf.: Wittgenstein 1969, 32.

15 Husserl's formula of intentionality is: "...nur Eines ist präsent, das intentionale Erlebnis, dessen wesentlicher Charakter eben die bezügliche Intention ist." Husserl 1968, 372. Also, only one thing is present, the intentional experience, whose essential descriptive character is the intention in question.

between the ink in my notebook and Mr N? There is no apparent relation between my letters and Mr N. Brentano argued that a sign as such does not relate to a specific object, but rather to the thought described by this sign.[16] According to this view the meaning is, as it were, behind my signs or letters and refers to a process. Could we make this process visible, so that the reference to Mr N only manifests itself on the mental level but not in the letters written in ink. It would show something that cannot be read out of the shape of my letters, namely the reference to the suspect Mr N. Unfortunately, this statement is untenable. While listening to a word the internal processes vary according to which word I hear but a supposed relation between an internal process and a sign would be difficult to understand.[17]

When I hear N's name a visual image of N may emerge in my mind for a moment. But what kind of image is that? Does it correspond to the facial composite that I was shown regarding the crime? Does it perhaps remind me of the photograph of someone I know or does it provoke a mental image of the signature of the person I know? Can we establish an unequivocal relation between Mr N and the crime? If we assume an internal process we are confronted with the same problem of a definite ascription of the facial composite to the suspect. To sum up, the question is this: how can a piece of reality (e.g. a facial composite), a photograph or a person's signature refer to such real phenomena as a crime or a person? To conclude, the concepts of consciousness and mind seem inappropriate means to bridge the gap between image and reality.

The shortcomings of Brentano's proposal may be located in the following argument: when making the following statement while considering a facial component, a photograph or a signature: "This image depicts the criminal", "This photo shows Mr N", "This is Mr N's signature", I may be wrong because the image only reminds me or N. But in fact, N has not committed the crime. And if during my thinking of N a mental image of N is provoked, then I cannot ask the following question: "Am I probably mistaken? Or, "Have I just thought of a criminal? This line of thought leads to an impasse. The error is that one tends to associate the process of thinking with an emerging mental image that was retrieved from our memory.[18] One may associate a mental image with different objects, be it a facial composite or a photograph without being aware of a possible error. In fact, such an incorrect association is not easily detected.

16 Brentano 1995, 88 f. Brentano 2008, 106.

17 Padilla Gálvez 2016, 4 ff.

18 Padilla Gálvez 2016, 62 f.

Brentano would respond that the representation is not the only criterion on which an attribution of meaning rests. Apart from the mental image there is an additional element that produces intention. But what is this element? It seems that Avicenna's notion of the "floating man" has serves as a model for a link between mental representation and object.

3 Brentano's Puzzle of Intentionality

The following three propositions have recently been suggested by proponents of intentionality, of which the first seems most relevant: the proposition of object-directedness.[19] According to this view, mental actions are immediately directed towards objects though actions and objects are not identical. According to the second proposition, mental inexistence is a special kind of existence (*besondere Art des Seins*) to perceive objects adequately. Brentano called this "*esse intentional*" or "*intentional inexistence*"[20].

According to the third proposition intentionality is a distinctive feature of consciousness,[21] in the sense that only conscious states are intentional. Intentionality is often associated with an additional fourth feature which is the knowledge of non-existence. The question of how one can recognize non-existing objects in the world can be answered by applying the concept of intentional existence. The argument is based on Avicenna's thought experiment of the 'floating man'. Would such a being ever be capable of thinking? The answer is affirmative, since this person is conscious of himself. Although the question and its answer reveal the absurdity of this view, the magical conception of intentionality of consciousness seems to have prevailed until nowadays.

A simple observation puts the three propositions in doubt if it were not for a few words that we frequently use in everyday language: "not", "and", "if", "well", "all".[22] None of these words fits into the scheme mentioned above since they lack a referential object.[23] We have seen that a simple negation prevents us

19 Brentano 1995, Book II, chapter 1, sect. 5. Brentano 2008, 106 ff.

20 The intentional inexistence is "... those phenomena which contain an object intentionally within themselves." Brentano 1995, 89. Brentano 2008, 107.

21 Intentionality "... is characteristic exclusively of mental phenomena." Brentano 1995, 89. "Diese intentionale Inexistenz ist den psychischen Phänomenen ausschliesslich eigentümlich." Brentano 2008, 107.

22 Wittgenstein 2001, §§ 178–183.

23 Wittgenstein 2003, 448 f.

from clearly determining an object. The use of “not” makes it impossible to establish a relationship between language, action and consciousness.[24] Let us consider another thought experiment: some argue that the meaning of a word is a mental act to clearly associate an object with a word. One could make the following objection and ask us to consider the meaning of the following requests: “Pronounce the word ‘not’ and mentally retrieve its meaning. What are you thinking of? What is the intention associated with “not”? One could name some model examples of statements in which the word ‘not’ occurs: “Do not eat”, “I do not want to give a lecture”, or “Do not drink water unless it is clean!” These and other similar examples reveal that the meaning of “no” or “not” is not associated with any specific mental state.[25]

So what does the thought of the word ‘not’ consist of? This question seems contradictory as it is implicitly assumed that thinking is a process or similar to a procedure. The notion of “thinking of a reference” raises new questions, such as which criteria this process follows. It seems difficult to describe these criteria. The words ‘no’ or ‘not’ appear in quite different contexts. We are not interested here in a process in which somebody recites a word in parrot-fashion. The words ‘no’ or ‘not’ normally describe situations in which somebody makes a mistake, becomes aware of it and consequently corrects it. So which processes are involved when using a negation? In other words, can we give a rule according to which these words are used correctly?

We do not need to examine our consciousness to answer this question. Nor do we have to determine the object-directedness of the word. No intention is associated with the word ‘not’. One is simply asked to show the uses of the word in different contexts. It is about demonstrating how one uses negation correctly.

4 Which Psychological Processes Are Involved?

Coming back to the original question of what it means to understand an action, I take the view that an action does not involve any additional psychological processes apart from the perception of the course of actions. When I carry out an action or happen to perceive one, various processes are going on inside me. What happens is that a mental picture emerges that is associated with further related pictures. Perhaps these mental images are even interfused with a feeling

24 Padilla Gálvez 2016, 257 ff.

25 Wittgenstein 2001, § 304.

tone. But all these associations fail to provide an answer to our initial question. What are characteristic elements of human actions? The answer is that one comprehends an action while applying it.[26] One starts an action because one wants to operate something. And we comprehend what the other is doing only if we are familiar with these kinds of actions and are able to operate in line with the other. Wittgenstein puts it like this:

> To intend a picture to be the portrait of so-and-so (on the part of the painter, e.g.) is neither a particular state of mind nor a particular mental process. But there are a great many combinations of actions and states of mind which we should call "intending".[27]

He refuted the view that understanding an action is comparable to sensations of pain, such as for instance, a person suffering of toothache.[28] And what is more, we aim to show that action has nothing to do with mental state. If we ask someone "Do you comprehend the meaning of the word 'drink'?", the expected answer is something like "Yes, it means to take something liquid or consume a drink."[29] In this explanation we do not presuppose an intrinsic knowledge of the word "drink". If someone had such knowledge one would automatically know the meaning of "*beber*" and "*trinken*" too. One would have memorized the words in one's mother tongue and would have simply forgotten their meaning the foreign language. We could equally assume that a person who had been working all his life but does not know that this action is called "work" meaning "to perform activities".

> There are and endless variety of actions and words, having a family likeness to each other, which we call "trying to copy.[30]

I can only say I am aware of the meaning of "drink" in exactly the same way as I know that the root of four is two, namely, not in the form of a state but in the form of a disposition.[31] We can only comprehend an action if we have learned how to perform it. Comprehension is only achieved when one was taught to perform an activity in a particular way. Suppose a master tells his apprentice to bring him a particular hammer to drive a nail into the wall. The instruction con-

26 Wittgenstein 2009, §§ 185–242.
27 Wittgenstein 1969, 32.
28 Wittgenstein 2009, §§ 243–315.
29 Wittgenstein 2009, § 150.
30 Wittgenstein 1969, 33.
31 Wittgenstein 1982, § 178.

tains the word "hammer"[32] as part of a process of sequential operations. This includes not only the ostensive definition of the word but also the execution of a related command. Why do we use a word for an action? Wittgenstein reminded us that enunciating a word goes hand in hand with a corresponding action. Language acquisition is never an isolated event but part of an established sequence of actions. The use of such words as "work", "hammer" or "drink", etc. always goes along with a particular activity.

5 Intentionality as Problem

Finally we shall address the paradox of intentionality which can occur in different forms. We try to imagine the following situation: the artist painter Diego Velázquez was contracted to paint a picture and expresses his intention accordingly:

> I will paint the five-year-old princess Margarita surrounded by her chambermaids.

An amateur painter who has seen Diego Velázques' picture '*Las Meninas*' and feels inspired by it says this:

> I will also paint the picture '*Las Meninas*'.

What is the difference between the artist's statement and the amateur painter's statement? Which intention does each of them have? What does the amateur painter have in mind when expressing his intention? These questions are to be solved.[33]

A painter has announced his intention to paint a picture. It is true that the meaning of "paint" can be interpreted in different ways. The listener of this statement must be aware that neither the theme and technique nor the painter's skills are considered. At least we can say that he intends to paint a picture and thereby creates a reinterpretation of this projection.[34] According to Wittgenstein, it is typical for intentional expressions that they are often opaque. The recipient

---

32 Wittgenstein say: "Suppose someone said, "*All* tools serve to modify something. So, a hammer modifies the position of a nail, a saw the shape of a board, and so on."" Wittgenstein 2009, § 14.

33 Foucault 1966, Searle 1980, 477.

34 Wittgenstein 1969, 33.

of such ambiguous statement may be tempted to over-interpret it and could draw wrong conclusions.[35]

Instead a clarification of ambiguousness is needed. Therefore the amateur painter has to clarify his intention to paint '*Las Meninas*' more precisely. The amateur painter then starts with his brushstrokes on the canvas without any artistic skills and technical knowledge. Due to his lack of skills the copying proves more difficult than expected. The final result is a copy consisting of brushstrokes. It has some similarity with the original but does not produce the same effect in the spectator. What has happened? An amateur painter is unable to distinguish between brushstrokes and the application of refined painting techniques, such as accentuation or perspective-related painting.

Perhaps it is the amateur painter's intention to show that a picture can easily be created by applying simple brushstrokes. But what he actually does is trivialize the whole concept of painting. But how can we distinguish intentional expressions from those hiding a different intention? In order to answer these questions we have to look into the following paradox: when expressing an intention the speaker must be able to explain why he has this particular intention.[36] Why does Velazquez want to paint a picture of Infanta Margarita? Why does an amateur painter want to paint the same picture? The answer lies in the meaning that each of them associates with the word 'paint'. If we are only interested in the reason for painting we forget to ask who actually has the necessary know-how to produce this picture. Each painter's statement is located in a different frame. If we tacitly assume that both associate the same skills when talking of painting we are misled. We shall demonstrate that by the implications following from this short dialogue:

– Why did you kill him?
– He killed my father.[37]

The interaction is arranged as an ideal question–response setting which does not clearly reveal the agent's reasons. When reading the dialogue we are automatically tempted to assume that the motive for the murder was revenge.[38] This model does not permit another solution, such as for instance, a third person being the murderer. Or what is more, the murderer may have arranged the evidence in such a way that the victim's son appears guilty. On the contrary, if we

35 Wittgenstein 1969, 35.
36 Anscombe 1963, 9.
37 Anscombe 1963, 10.
38 Wittgenstein 1982, § 175.

ask for the reason we tend to follow a fixed pattern, that of revenge in our case. Other scenarios are possible, for instance the victim might have been killed accidentally or by anybody whom he had not known before. But all these possibilities are cleared out without being even considered.

Wittgenstein provided a detailed analysis of questions regarding rule-following which he called "*knowing how to go on*?"[39]. If we apply the principle of following a rule to our examples we may ask what actions must be carried out in order to fulfill an intentional statement in the future. Suppose that there is a third possible committer and we portray the circumstances in such a way that all the evidence points to the son. As the revenge motive seems obvious to the policeman, the wrong person might be incarcerated and the case is closed. However, if we ask for the procedure how the murder happened, we may detect the details and the committer of the crime. If the alleged committer claims to be a pacifist we are confronted with a paradox. Wittgenstein held that such paradoxes can only be solved by asking for both, the reason and the procedure. The answers to both questions should then be contrasted and should be free of contradiction. As such, we have shown that the solution to the paradox of intentionality is much more complex than just a simple search for obvious reasons.

6 The Paradox of Intentionality

Intention can be characterized by three aspects: let defined "O_I" be what Brentano calls "mental inexistence of an object", "R_I" stands for "reference to a content" and "D_I" is "distinction to an object", also described as "immanent or internal objectivity".[40] If we use these three aspects to describe intentionality (I) we receive the following model:

(I) $I{:}\langle O_I, R_I, D_I \rangle$

On the other hand, we have three corresponding aspects that denote the occurrence of the intended object. Let "O_E" be the "existence of an object", "R_E" the "reference to the content" and "D_E" stand for external objectivity. Accordingly, the fulfilment (F) of an intention can be described by this formula:

(F) $F{:}\langle O_E, R_E, D_E \rangle$

39 Wittgenstein 1969, 40.
40 Brentano 1995, 88 f. Brentano 2008, 106.

If a speaker intends to realize a non-existent object by a establishing a relation to content being directed towards an object, the fulfilment of this intention is not necessarily guaranteed. The preconditions allow two possible results: either the speaker's intention is accomplished or it is not. This paradox corresponds to the following scheme:

(1) If S directs his intention to the existence, relation and direction to an object and all these three criteria are met, then the intention is fulfilled.
(2) If S directs his intention to the existence, relation and direction to an object and all these three criteria remain unsatisfied, then the intention remains unfulfilled.

This can be described in the following form:

[I_1] If (S has the intention that $<O_I, R_I, D_I>$, and S satisfies $<O_E, R_E, D_E>$), then the intention is fulfilled.
[I_2] If (S has the intention that $<O_I, R_I, D_I>$, und S does not satisfy $<O_E, R_E, D_E>$), then the intention remains unfulfilled.

This leads us to the following simplified form:

(1) $I{:}<O_I, R_I, D_I> \rightarrow F{:}<O_E, R_E, D_E>$
(2) $I{:}<O_I, R_I, D_I> \rightarrow \neg F{:}<O_E, R_E, D_E>$

We shall illustrate what was described by an example. Suppose Maria strives to become a professional singer and does everything to achieve her objective.

(1*) If Maria has the intention to become a singer and studies singing, then she will achieve the desired career.

Nevertheless, there is no guarantee that automatically assures this result. What we have is a Ciceronian implication[41] and not a strict implication and therefore its fulfilment depends on several aspects. On the contrary, we may propose this:

(2*) If Maria has the intention to become a singer but does not sing, then her desire cannot be fulfilled.

Generally, there is no guarantee that Maria's intention to become a singer will be fulfilled in the future.[42] The conjunction of conditions in [I_1] und [I_2] is as obvious as the bi-conditional scheme of de-nominalization and thus does not

41 Cicero 1880, *Academia priora*, XXIX, 95.
42 Wittgenstein 1982, § 6.

require recourse to the speaker. The intentional conditionals in (1) and (2) show that when applying these principles we refer to a speaker who says that he intends something unfulfillable and thereby creates paradoxes. This implies that S intends to express something fulfilled when S intends at the same time to get something unfulfilled.

7 Conclusion

The approach outlined in this article produces peculiar consequences. Which implications are associated with the meaning of the word 'not'? How can we explain the meaning of negation? Suppose an apprentice is told *not* to bring a specific hammer to his master. What should he do then? Should he bring another hammer or none at all? Can the word "not" help establish a relation between language, action and thinking (consciousness)? Is the meaning of the enunciation and a corresponding action determined by a particular experience? The apprentice was told not to bring a particular. If the master teaches the apprentice to perform an action he shows him how to apply the word hammer by demonstrating how to use different hammers for different materials. The apprentice learns to use a special hammer to drive a nail in the wall. Consequently the apprentice will not bring a mallet nor a gavel but rather a claw hammer. As such, the whole meaning of the word hammer must be included in its correct use in a situation.

At first sight, action appears as a process which is accompanied by certain words. But this view is mistaken. It is assumed that action is a process that develops with time. Action is not a particular mental process that adds up to the perception of the proposition. What is of interest for the master and his apprentice is the disposition of each hammer and consequently its actual function.[43] To use the word correctly implies to know the function of the instrument and how to apply it in order to fulfill its function. One would not use a sledge hammer with fragile materials because of its destructive effect. A disposition goes along with the correct use of words that point to a particular action.

Action is not a process in progress. Still, it is not completely timeless and unrelated to actual processes in our consciousness. To carry out an action implies knowing how to apply an instrument. It is not a process but rather consists of several acts that expand over time and space and require repeated modifica-

43 Wittgenstein 1982, § 178.

tion and adaptation. Just the same as in the case of the apprentice, it is a skill that unfolds over time.

8 References

Anscombe, G. E. M., 1963, *Intention*, Oxford, Basil Blackwell.

Avicenna (Ibn Sina), *Kitab al-Isharat wa-'l-tanbihat* (*Remarks and Admonitions*), ed. S. Dunya, Cairo, 1960 (parts translated by S.C. Inati, *Remarks and Admonitions*, Part One: Logic, Toronto, Ont.: Pontifical Institute for Mediaeval Studies, 1984.

Avicenna (Ibn Sina), 1968, Avicenna Latinus, *Liber de anima, seu sextus de naturalibus,* IV - V. Eds. S. Van Riet, G. Verbeke. Louvain, Éditions Orientalistes E.J. Brill.

Avicenna (Ibn Sina), 1972, Avicenna Latinus, *Liber de anima seu sextus de naturalibus* (*Kitāb aš-Šifā'*), I, II, III, ed. S. Van Riet, G. Verbeke. Louvain, Peeters.

Brentano, F., 1995, *Psychology from an Empirical Standpoint*, ed. L. McAlister. London, Routledge.

Brentano, F., 2008, *Psychologie vom empirischen Standpunkt*. Frankfurt a.M., Ontos Verlag.

Cicero, M. T., 1880, *Academicae quaestiones*, in: Cicero, *Opera Omnia*, Vol. VIII. Leipzig, G. Teubner.

Descartes, R., 1964-1974, *Meditationes de prima philosophia, in qua Dei existentia et animæ immortalitas demonstrator*, in: *Oeuvres de Descartes* edited by Charles Adam and Paul Tannery, Paris, Vrin-CNRS.

Foucault, M., 1966, *Les Mots et les Choses. Une archéologie des sciences humaines*. Paris, Gallimard.

Gaffal, Margit, 2015: Forms of Life as a Philosophical Concept. In: *Emergent Forms of Life in Anglophone Literature. Conceptual Frameworks and Critical Analyses*, M. Basseler, D. Hartley, A. Nünning (eds.), Wissenschaftlicher Verlag, Trier, 27-37.

Gundissalinus, D., 1999, *De Processione Mundi*. Eds. M. J. Soto Bruna, C. Alonso del Real. Pamplona: Eunsa, 1999, p. 152; 23,21f.

Gundissalinus, D., 2009, *Tractatus de Anima*. Ed. M. J. Soto Bruna. Pamplona: Eunsa.

Gyekye, K., 1971, The Terms 'Prima Intentio' and 'Secunda Intentio' in Arabic Logic, *Speculum*, 46, 32-38.

Husserl, E., 1968, *Logische Untersuchungen. Untersuchungen zur Phänomenologie und Theorie der Erkenntnis*, II/1. Tübingen, Max Niemeyer.

Marques, António and Nuno Venturinha (eds.), 2010: *Wittgenstein on Forms of Life and the Nature of Experience*. Peter Lang, Frankfurt a.M.

Padilla Gálvez, Jesús, 2010: Form of Life as Arithmetical Experiment. In: Antonio Marques and Nuno Venturinha, 2010, 113-124.

Padilla Gálvez, Jesús, 2011: Language as Forms of Life. In: Jesús Padilla Gálvez und Margit Gaffal (eds.), 2011, 37-56.

Padilla Gálvez, Jesús, 1016: Lebensform und Sprache. In: *Lebenswelt und Lebensform*, Phänomenologische Forschungen 2016, (Eds.) Christian Bermes, et allii., Felix Meiner Verlag, Hamburg, 257-273

Padilla Gálvez, Jesús (ed.), 2010: *Philosophical anthropology: Wittgenstein's Perspective*, Ontos Verlag, Frankfurt a. M.

Padilla Gálvez, J. (ed.), 2016, Action, Decision-Making and Forms of Life. Berlin, Boston, Walter de Gruyter Verlag.
Padilla Gálvez, Jesús and Margit Gaffal (eds.) 2011, *Forms of Life and Language Games*, Ontos Verlag, Frankfurt a. M.
Searle, J. R., 1980, Las Meninas and the Paradoxes of Pictorial Representation, *Critical Inquiry*, Vol. 6, No. 3, 477-488.
Searle, J. R., 1983, *Intentionality. An Essay in the Philosophy of Mind*. Cambridge, Cambridge University Press.
Thomae Aquinatis, 1948, *Summa contra Gentiles*. Edition Leonina manualis. Torino, Marietti.
Wittgenstein, L., 1969: *The Blue and Brown Books*. Oxford, Blackwell Publishing.
Wittgenstein, L., 1982: *Bemerkungen über die Philosophie der Psychologie*. Vol. 2, eds. G. H. Von Wright and H. Nyman. Frankfurt a. M., Suhrkamp.
Wittgenstein, L., 2001: *Philosophische Untersuchungen. Kritisch-genetische Edition*. Ed. J. Schulte in collaboration with H. Nyman, E. von Savigny and G. H. von Wright. Frankfurt am Main, Suhrkamp.
Wittgenstein, L., 2003, *Ludwig Wittgenstein and Friedrich Waismann: The Voices of Wittgenstein. The Vienna Circle*. Ed. Gordon Baker, tr. Gordon Baker, Michael Makkert, John Connolly and Vasilis Politis. London and New York, Routledge.
Wittgenstein, L., 2009, *Philosophische Untersuchungen / Philosophical Investigations*. Transl. by G.E.M. Anscombe, P. M. S. Hacker and J. Schulte. Revised fourth edition by P. M. S. Hacker and J. Schulte. Oxford, Wiley Blackwell.

Michel Le Du

Intentionality and Monitoring

1 Action Disorders

As everyone knows, the word “intentionality is used to mention a feature of actions as well as a property of mind and the use of the same term in both contexts cannot be a mere coincidence. However, my goal in the present paper is not to explore these conceptual connexions. I want to focus on the intentionality of action and examine some of the empirical treatments it aroused. I do not want to suggest simply that psychological or even experimental approaches to intentionality are illegitimate as such. But the fact is that they are infected by several conceptual confusions to such a point that one sometimes gets the feeling that intentionality is not what these approaches really deal with.

During the last decades, many efforts have been dedicated to the scientific study of dysfunctions affecting agency.[1] One of the most striking examples of such troubles is the thought insertion syndrome, marked out by the agent’s propensity to describe himself as suffering from thought intrusion: the patient tells the doctor that he hears voices in his head and is apparently unable to see that what he describes as an intrusion in his own mind is indeed his own inner talk. This syndrome is characteristic of schizophrenia. As we can see, its most astonishing feature is the discrepancy between an action initiated by the agent, on the one hand and the comment he delivers, on the other hand: he describes his action in impersonal terms, as if it was something foreign to his will, something simply “happening” to him.[2]

Such cases do not simply confront us with bursts of delirium, although many of those patients interpret the voices as the result of a manipulation by the Martians or some CIA office: the point is that they really *experience* these voices as alien and the main problem is with explaining this experiential dimension. A widespread interpretation among cognitive scientists and psychiatrists versed in the study of schizophrenia consists in saying that people enduring this trouble fail to achieve *self-monitoring*. As everyone knows, a television

1 See for instance Frith 1992.

2 A similar discrepancy is a characteristic feature of another syndrome, for which the paradoxical term *blind vision* has been coined. Weiskrantz 1986 and Weiskrantz 1997. For an astute philosophical study of these issues: Hyman 1991, 166–200.

https://doi.org/10.1515/9783110560282-089

monitor is a screen through which on-going events can be scrutinized and controlled. And the word "monitoring" has already been used (even in French) for a long time to denote a surveillance technique. According to the authors I alluded to, monitoring provides the agents with a sense of their own agency. And its failure results in the agent experiencing actions of his as commanded by an alien force. It's important that we evaluate the capacity of this notion to shed light on normal agency as well as on the ordinary flaws in action control. If, indeed, the monitoring hypothesis reveals its flimsiness in such cases, there will be good reasons to think that it cannot account either for the kind of radical breakings we mentioned a minute ago.

2 Monitoring and Egocentricity

Let us pick up an example from a paper by Ian Gold and Jakob Hohwy.

> Suppose one of my goals is to go to work on time each morning. On a particular morning, I form the intention to catch a bus and then perform the action of thinking: "Catch the bus!" If I have not monitored the intention to catch the bus, I will find myself with a particular thought without any awareness of the intention that initiated it. As a result, I may experience the thought as having been put into my head by someone else – a delusion of thought insertion. If the thought "Catch the bus!" occurs as inner speech ... I will experience it in the form of a hallucination in which someone is *telling me* to catch the bus.[3]

There are many questionable things in this explanation. First, describing thought as an action ("the action of thinking") is mistaken. Although reasoning and inferring can be described as things we do, the verb "to think" doesn't denote actions for the most part: thoughts, as we all know, come and go.[4] Secondly, it's wrong to suggest that I have to form the intention to catch the bus beforehand, in order to think: "Catch the bus!" and eventually to catch it. Someone may get on the bus intentionally without having previously formed the intention to do so.[5] Thirdly, suggesting that the explanation of a thought consists in exhibiting the intention it comes from is equally wrong. In most cases, when a strange thought pops up in one's mind, explaining it doesn't consist in connecting it to a previous intention, but in seeing it in a different perspective. But there are two more serious sources of trouble.

---

3 Gold and Hohwy 2000, 145-165 and especially, 150–151.

4 Nietzsche 2015, § 17.

5 Anscombe 2000 (1957).

(1) Having a feeling of strangeness towards a thought is one thing; regarding it as the result of thought insertion is quite another thing. Although I occasionally find some of my own thoughts puzzling, to the point of asking myself how exactly I came to think them, I never consider them alien and never think they are not *mine*. Consequently, there is a gap between regarding one's thoughts as strange and regarding thoughts occurring in one's mind as alien results of thought insertion. The last process is not the first one brought to a higher degree: it's different in nature.

(2) According to the text we have just quoted, our intentions are destined to a grasp by *self-monitoring* and self-monitoring is supposed to provide action with egocentricity. Egocentricity is precisely what the patient with thought insertion syndrome lacks. It's worth noticing that the concept of monitoring has a definite application in connection with certain kinds of artefacts and its use as a technical word for quality controls is attested as early as 1924, according to the New English Dictionary. When quality controls are carried out, items are subjected to a procedure called *template matching*: the parts are monitored in order to see if they fit an image built into the machine. So it's not surprising to see that when used to describe self-awareness, the word suggests a capacity modelled on perception. And my reservations are about the monitoring explanation of self-awareness, note about the concept of monitoring as such, as this explanation looks much like an updated version of the traditional idea of an inner sense. But does it make sense to think that intentions fall within the scope of such an internal apparatus? Can we say that we sense our own volitions and intentions? Anthony Kenny writes:

> If volitions were genuine mental events occurring with frequency ... it should be possible for any articulate human being to answer questions about the nature, occurrence, timing intensity and qualities. But this is notoriously not the case.[6]

As we can see, the problem with this account is not only that one can be suspicious of the very notion of an inner sense, but that even if such a faculty did exist, intentions would not belong to the category of things falling within its scope: we do not feel them inwardly. In other words, I do not see any difficulty in describing the kind of action disorder I mentioned earlier in terms of a problem with egocentricity but, conversely, I think that the self-monitoring account of egocentricity is doomed by confusion. This is the reason why I'm equally

6 Kenny 1989, 32.

suspicious of the egocentricity failures explanation in terms of monitoring faults.

3 Intentionality and the Will

In fact, there is even an additional unclarity, as it is not easy to see how the monitoring hypothesis applies to the thought insertion syndrome. Brooding over something or scratching one's head can surely be called things people *do*. These are generally things one doesn't do on purpose but that one can inhibit with an effort. And although someone scratching his head is normally able to give an answer when asked: "What are you doing?" this is not enough to consider his scratching an intentional action. Nevertheless, scratching one's head is different from moving my arm after a doctor has put electrodes on my skull to test my reactions.

Accordingly, the distinction between *intentional* and *voluntary* actions should not be overlooked. Many voluntary actions are not intentional and talking in one's head is rather a voluntary action than an intentional one. Someone who speaks inwardly may try to stop his mental discourse, in order to focus his attention on something else, for instance. This is something the schizophrenic mentioned earlier cannot achieve. But if indeed an ordinary agent can normally control his inward speech through the alleged monitoring process, such a monitoring seems to provide him, in that case, with a sense of his inner speech's voluntariness, but not of its intentionality, strictly speaking. Anyway, I think that the idea of an inner sense of one's actions voluntariness and the idea of an inner sense of one's actions intentionality are both confused. An intention is not an episode of one's mental life, but the voluntariness of an action is not something that occurs in one's mind either. And this suggests that the very notion of voluntariness deserves a closer examination.

Anthony Kenny makes a distinction between two kinds of *wants*.[7] He contrasts what he calls "animal wants" or "desire" with "volitions". "Animal wants" correspond to what classical Greek philosophers called *epithumia* and scholastic *concupiscentia*. Desire, unlike volition, seeks immediate satisfaction, as it is a want for something now. In other words, it is a want that is felt more or less continuously until it is satisfied. Volition, on the contrary, may be for something distant in time, and, he adds, may be operative "without being an item in

7 Kenny 1989, 35–37.

the flow of consciousness".[8] He also insists that "the kind of behaviour which manifests desire is voluntary action" and that "the kind of behaviour which paradigmatically manifests volition is intentional action".[9] There is a certain degree of regimentation in terming such a distinction in that way. On the one hand, the word "desire" is, in fact, often also used to name a relation to something remote in time; on the other hand, our wants in need of immediate satisfaction can be either natural appetites (like thirst or hunger) or mental states directed to more specific objects. But this is a minor point, as the main justification for the distinction is the idea (1) that acting intentionally is acting for *reasons* (2) the relation between reason and action is *logical*. This is the ground for saying that volitions can be operative without belonging to the flow of consciousness. Anthony Kenny says:

> Intentional actions are actions which are chosen either as ends in themselves or as means to some other ends: if actions are chosen for their own sake, then not only are they intentional but they may be called purposes.[10]

This is his ground for saying that, although many animals have wants, they do not have volitions. One can feel a desire (in Kenny's sense) and act out of this desire, but this doesn't mean that one senses inwardly the voluntariness of one's action as such. In fact, the adjective "voluntary" is at its place when the action is *controlled*. Traumatic post-images, for instance, are a normal aftermath of painful experiences. Trying to preclude such images' recurrence is a frequent reaction. In this sense, these images can be called voluntary (On the contrary, exerting one's will in order to change the perceptual objects one sees, as Wittgenstein remarks, doesn't make sense). But I cannot see on which empirical grounds the idea that such a control is accomplished through a self-monitoring apparatus can be backed up. In fact, it looks like a philosophical speculation disguised as an empirical claim.

Accordingly, my conclusion so far is twofold. (1) Thinking that intentions can be subjected to self-monitoring is mistaken, as they do not have duration and are not ingredients of one's mental life (contrary to mental images, emotions or occurring thoughts). (2) As far as a trouble with agency, such as thought insertion, is concerned, describing this trouble as a failure at controlling one's intention is equally mistaken: speaking inwardly falls under the category of voluntary actions, not of intentional actions. However, as one can guess from

8 Kenny 1989, 36.
9 Kenny 1989, 37.
10 Kenny 1989, 45.

the previous remarks, I do not think that something like spotting the voluntariness of an action through a monitoring device does exist either. Let us now return to thought insertion for a while.

4 Experiential Irrationality

In the paper I mentioned earlier, Gold and Hohwy try to account for the experiential dimension of this syndrome. Their idea is that in order to explain such a delusion as well as kindred phenomena, a new department of irrationality should be introduced, for which they coin the name *experiential irrationality*. They spend much time refuting in detail other explanations. They contend that thought insertion delusion cannot be interpreted as a violation of *procedural rationality*. Such a violation occurs when a principle of reasoning is broken. Along these lines, someone who believes in thought insertion can be described as violating the principle of *reflective equilibrium*, which states that we should aim for maximal coherence among our beliefs.[11] But, as they note:

> An inability to achieve reflective equilibrium among beliefs would infect the agent's whole system of beliefs, and one would expect the agent to experiment delusions of every imaginable kind.[12]

I think this is quite right. Such an explanation is unable to account for the fact that thought insertion delusion is circumscribed: the agent is able to achieve reflective equilibrium regarding most of his other beliefs.

Next to the procedural account of irrationality, is the *content account*. According to it, a delusion will be said to be irrational if "it assigns ineligible, unreasonable content when a more eligible assignment would have fit behaviour equally well."[13] As the two authors notice, the content account doesn't do any better than the explanation in terms of break in reflective equilibrium. They conclude quite correctly that:

> The attempt to reduce the irrationality of delusion to the bizarre contents of delusional beliefs suffers from the same problem as the procedural account just evaluated: a content

11 Gold and Hohwy 2000, 154.
12 Gold and Hohwy 2000, 156.
13 Lewis 1986, 155.

> account does not explain why the bizarre beliefs of these subjects seem so narrowly restricted and predictable in content.[14]

As these two accounts fail to do the job, hypothesizing another sort of irrationality seems to be the reasonable option. Here is how they introduce this option:

> The source of thought insertion ... is the experience itself of the schizophrenic subject, and, in particular, its alien quality. The elaboration of the delusion in hypotheses and ancillary beliefs should be understood to be derivative from, or secondary to, this experience ... At least some delusions, therefore, are best explained as *disorders of experience* rather than disorders of belief, desire or reasoning.[15]

I think that there is nothing objectionable with this idea of *disorders of experience*. But, again, I consider that the self-monitoring account of these disorders is confused. To repeat myself, objects of perception, unlike imaginary objects, cannot be changed at will and the word "monitoring" suggests precisely an apparatus modelled on perception. Accordingly, monitoring includes the eventuality of noticing or spotting some unexpected event and this is exactly what the schizophrenic hearing inner voices with an "alien quality" thinks he does. So it seems that, up to a certain point, the word better fits the egocentricity failure than the normal situation. Consequently, my conclusion so far is that although the notion of a self-monitoring is definitely confused, it's acceptable to use the word *monitoring* (without the prefix "self") to describe the schizophrenic patient's attitude towards inner voices, as he considers them will-independent (as perceptual objects are). Accordingly, I more or less reverse the standard neurocognitive account. But this is, at least partly, a terminological issue.

5 Feelings

One of the purposes of the previous sections was to show that it is wrong to think we can feel inwardly the voluntariness or the intentionality of our actions. However, the verb "feel" has two different senses.[16] First, I can say that I feel a lump in the mattress. In such a case, the verb refers to a *sensation*. But it is also used to mention affective states like emotions. Can't we say then, although we do not have an inner sense, providing us with an access to our actions' inten-

14 Lewis 1986, 158.

15 Gold and Hohwy 2000, 160.

16 Kenny 1989, 53.

tionality, that these actions are accompanied by a distinctive feeling? Several writers have made this suggestion and it can be traced back to William James' *Principles of Psychology*.[17] The problem is that acting can be accompanied by various feelings, none of them being characteristic of acting as such. Accordingly, the quest for a feeling specific of acting seems hopeless.

This point is clearly made by John Searle in his well-known book *Intentionality*, but I think that Searle's own treatment contains several unclarities.[18] One thing, however, is clear: the notion of an "experience of acting" he introduces plays the role fulfilled by the monitoring hypothesis in the neurocognitive theories we mentioned earlier. Searle rejects explicitly the idea of "a characteristic experience common to every intentional action" but contends at the same time that there is "an obvious phenomenal difference between the case where one moves one's hand and the case when one observes it move independently of one's intention".[19] The two cases, he adds, "feel different to the patient". At first sight, it's difficult to understand how such an obvious phenomenal difference can be felt if no characteristic experience common to all intentional action does exist. Searle would probably answer that a locution such as *experience of* can be understood in two different ways. When I see a donkey, I have a perceptual experience *of* this donkey and this mean that the donkey is what my experience is *about* (or *directed to*). In that sense, there is no distinctive experience common to all intentional action. But we also speak of the experience of swimming and such an experience is not about swimming: up to a point, swimming and experiencing swimming, are one and the same thing. And, strictly speaking, this experience cannot be named a feeling: experiencing swimming and feeling that one is swimming are two different things. In this sense, Searle explains, we may consider performing an action an experience. If he is right, the hallucinated patient we mentioned earlier is deprived of such an experience.

It is true that if a patient first moves his hand and then has his hand moved, as part of an experience led by a scientist, the two episodes feels different to him. In such a case, it makes sense to speak of a phenomenal difference. But this is not a sufficient as a ground for the general idea of an experience of acting. Normally, an experience can be contrasted with another experience: we can contrast the experience of swimming with the experience of climbing. But the alleged experience of acting cannot be contrasted with another experience in that way. No doubt intentional actions have distinctive features. But thinking of

17 James 1950, 492–493 and Russell 1995, chap. XIV.

18 Searle 1983.

19 Searle 1983, 90.

these features as actualities (sensations, feelings, and experiences) related to the conscious mind is mistaken. As Wittgenstein famously said, "Only you can know if you had that intention".[20] If an agent acts with an intention, has a prior intention or simply acts intentionally, he can normally declare his intention or the reason for his deed. Expressing one's intentions and motives boils down to exercising a power. This exercise doesn't require a self-monitoring device or an experience of acting, although acting can be accompanied by various experiences and feelings.

6 Intentionality and Proto-intentionality

An intentional explanation consists either in mentioning an intention, a motive in general (He did that out of love or admiration) or, as Elisabeth Anscombe says, a backward looking motive (like guilt, remorse).[21] At first sight, this account seems larger in scope than Kenny's one which was mentioned earlier. But, if an action is inspired by a backward looking motive (like revenge), it must consist in achieving such a purpose (=revenge on someone) or in being mean to such a purpose. And the same can be said of motives in general. As we have seen, animals can act voluntarily but ascribing them intentional actions strictly speaking is questionable, as they do not have linguistic abilities and cannot, consequently, put forward their purposes and motives. And, as the connection between intention and action is *logical*, feelings and experiences are only ancillary in that respect. In other words, it's difficult to ascribe intentional actions to creatures deprived of the concept of a purpose. Nevertheless, such creatures can accomplish *goal-directed* actions, actions that one cannot interpret without making use of the concept of a goal.

Even among human beings, many actions can be directed to a goal without involving the concept of a goal. Any developmental psychologist knows the actions of a suckling cannot, in most cases, be understood if one doesn't connect them to a purpose (which often consists in a beneficial state). As prelinguistic children don't have the concept of a goal, their actions can also be said *goal-directed* but not *involving the concept of a goal.*[22] These actions I call *proto-intentional.* But it's important that we notice that, as grown-ups, although

20 Wittgenstein 2000 (1953), § 247.

21 Anscombe 2000 (1957), § 13–16.

22 Perner 1993.

we have such a concept, we don't necessarily succeed in applying it to our deeds. In other words, we master the concept of a goal in general but sometimes fail to apply it to some of our own actions. This is typically what happens in a *Fehlleistung*. Such rather ordinary dysfunctions are one of the favourite topics of the psychopathology of everyday life and they are far from being as radical as the egocentricity failure we evoked earlier. The agent achieves a meaningful action whose meaning he doesn't see (although we may think that he should, as a rational person). Acting and not seeing that one's action has a meaning is different from mistaking acting for not acting.

7 References

Anscombe, E., 2000: *Intention* (1957). Cambridge, Harvard University Press.

Frith, Ch., 1992: *Cognitive Neuropsychology of Schizophrenia*. Milton Park, Taylor & Francis.

Gold, L. and J. Hohwy, 2000: Rationality and Schizophrenic Delusion, in: *Pathologies of Belief*, ed. Max Coltheart and Martin Davies, Oxford, Blackwell, pp. 145-165.

Hyman, J., 1991: Visual Experience and Blindsight, in: *Investigating Psychology*, ed. J. Hyman. London Routledge, 1991, pp. 166-200.

James, W., 1950: *Principles of Psychology*, vol. 2. New–York, Dover.

Kenny, A., 1989: *The Metaphysics of Mind*. Oxford, Oxford University Press.

Lewis, D., 1986: *On the Plurality of Words*. Oxford, Oxford University Press.

Nietzsche, F., 2015: *Beyond Good and Evil*. Ware, Wordsworth Editions.

Perner, J., 1993: *Understanding the Representational Mind*. Cambridge, MIT Press.

Russell, B., 1995: *Analysis of Mind*. London, Routledge.

Searle, J., 1983: *Intentionality*. Cambridge, Cambridge University Press.

Weiskrantz, L., 1986: *Blindsight, a Case Study and Implications*. Oxford, Clarendon Press.

Weiskrantz, L., 1997: *Consciousness Lost and Found*. Oxford, Oxford University Press.

Wittgenstein, L., 2000: *Philosophical Investigations* (1953). Oxford, Blackwell.

António Marques

Wittgenstein's Philosophy of Psychology on Intention

1 Introduction

In his introductory lines to the chapter on "Intending" of the magnificent volume four of his analytical commentary of Wittgenstein's *Philosophical Investigations* (*PI*), Peter Hacker notes that

> It is a surprising fact about the history of modern philosophy that a topic so central to philosophy of mind, ethics, jurisprudence, and indeed to the philosophy of mind as intending and the nature of intentional action was largely neglected"[1]. He also notes that the significant change came in the 50s largely under the stimulus of the *PI*, Gilbert Ryle's *The Concept of Mind* and some important literature in the field of philosophy of law, namely of the great philosopher of law H.L.A. Hart. In the words of Hacker "The obscure remarks on voluntariness and on intentions in *PI* 611-60 were a stimulus for a flood of writings which made the theory of action a central theme in contemporary philosophy[2].

As Peter Strawson remarked in one of the earliest reviews of the *Philosophical Investigations*

> ... rarely has a subject been treated so powerfully and suggestively in so few pages[3].

I intend to revisit some of these topics and I'll give special attention the sections of *PI* dedicated more directly to the concept of intention, which is one of central issue of the so called inner experience, I'm referring to the expressions of inner experiences like hoping, expecting, willing or intention. Expressions of intention and intentionality are often related with language-games of the same family of those of will. As we know the late Wittgenstein devoted a large part of his investigations to the language of psychological concepts, and among them intention and intentionality deserves a central place. In fact it is remarkable that in the last 30 sections of *PI* this is the main topic if not the unique theme. In fact these are topics already treated in the big hermeneutical Hacker's work, with

1 Hacker 1996, 619.
2 Hacker 1996, 619.
3 Quoted by Hacker 1996, ibidem.

https://doi.org/10.1515/9783110560282-099

great detail and deepness, and hope to remember and add something relevant. So in what follows I wish to justify the following claim: *philosophy of psychology carries out its therapeutic work on intention action, through analysis of intention language-games of past intentions and intentional predictions.* This means that the wittgensteinian analysis of intention expressions develops mainly in a double time axis.

2 Philosophy of Psychology

Let's start by making clear what it is meant here by philosophy of psychology. Philosophy of psychology in the wittgensteinian sense consists mainly in a therapy of various pictures of a metaphysical stance regarding psychological concepts. In fact I think that there is not a single metaphysical picture of inner life, that Wittgenstein has in mind and that more than one picture is at stake in the last sections of the *Philosophical Investigations*, as well as in its so called second part, and in the many reflections on philosophy of psychology that has been collected by G.E. M. Anscombe and G.H. von Wright[4]. Any way for our purposes now I'll refer only to the *Philosophical Investigations* sections from 571 on and to some passages of the so-called part two of that work.

What I referred to as metaphysical views of our inner life results mainly from the misuse of introspection attitudes that reveal themselves not to be the adequate method to explore the inner life. About introspection in *Philosophical Investigations* we found the following remark:

> Does it make sense to ask 'How do you know that you believe?' – and is the answer: 'I know it by introspection?' In *some* cases it will be possible to say such thing, in most not. It makes sense to ask: 'Do I really love her, or am I only pretending to myself? and the process of introspection is the calling up of memories; of imagined possible situations, and of the feelings that one have if…[5].

Relevant here is the role of memory in processes of the inner life. I'll mention it as a crucial element in the understanding of intentional language.

As Wittgenstein claims in *Philosophical Investigations* 656, having in mind the right method to explore de complexity of inner life, one has to look on the language-game as the *primary* thing…

4 Wittgenstein 1980.
5 Wittgenstein PI, § 587.

> And look on the feelings, etc., as you look on a way of regarding the language-game, as interpretation[6].

What he identifies here is a methodological point, by which it is through comparing, approaching and differentiating language-games that we can succeed to see clearer in the apparently super-organized inner life, structured by the use of either philosophy or even the common use of psychological language. It is a characteristic of the metaphysical picture that inner life seems to be filled out by a mental acts and that language expresses them in distinct moods. Does the philosopher hypostatize intention as Wittgenstein remembers regarding feelings in general? In fact he writes,

> When we do philosophy, we should like to hypostatize feelings where there are none[7].

Through introspection I wish I to capture this and that experience itself, for example that singular intention, which is the same under different tones. The opponent voice notes that it is completely different to state something with intention and without it. So how is it possible not to recognize the pure intention as the difference between both statements? The opponent voice of a philosopher speaks:

> But when you say 'I intend to go away' you surely mean it! Here again it just is the mental act of meaning that gives the sentence life. If you merely repeat the sentence after someone else, say in order to mock his way of speaking, then you say it without this act of meaning[8].

Wittgenstein answers that yes, this happens when we are doing philosophy. The philosopher supposes that he captures this pure element of intentionality when he compares a sentence without intention (for example I repeat without intention a sentence with intention, by mocking let's say). "But let us really think out various *different* situations and conversations, and the ways in which that sentence will be uttered in them" suggests Wittgenstein, so inviting the philosopher to transpose himself to the real life and its multiple language-games. – Anyway the opponent philosopher resists: "I always discover a mental undertone; perhaps not always the *same* one." To this Wittgenstein answers with two questions: "And was there no 'undertone' there when you repeated the sentence after someone else? And how is the 'undertone to be separated from the rest of

---

6 Wittgenstein PI, § 656.
7 Wittgenstein PI, § 598.
8 Wittgenstein PI, § 592.

the experience of speaking?". The same thing happens with language-games of will. The philosopher sees in the difference between "I raise my arm" and "the fact that my arm is raising" something that works as a *primum movens*. It is why

> One imagines the willing subject here as something without any mass (without any inertia); as a motor which has no inertia in itself to overcome. And so it is only mover, not moved. That is: One can say 'I will, but my body does not obey me' – but not: My will does not obey me" (Augustine). But in the sense in which I cannot fail to will, I cannot try to will either[9].

Will of course is something immediate and this quality can be formulated in following terms: I express my will, but I cannot describe it. "I raise my arm voluntarily" could just be a kind of statement that one would like to call a description of an act of will, since there is the possibility that I don't raise it voluntarily, for example as the result of an electric wave. But in this case *I put myself as observer* of the movement and can even show surprise or wonder. I raise my arm and I'm surprised is possible exclamation in the situation of the electric wave hitting me, but certainly it doesn't belong to an expressive language-game like "I raise my arm in order to call your attention". In *Philosophical Investigations* 627 Wittgenstein explores the difference between those language-games, where expression and description of the same bodily movement is at stake.

> Examine the following description of a voluntary action: "I form the decision to pull the bell at 5 o'clock, and when it strikes 5, my arm makes this movement"- Is that the correct description, and not *this* one: "... and when strikes 5, I raise my arm'? – One would like to supplement the first description: "and see! My arm goes up when it strikes 5'. And this 'and I see!' is precisely what doesn't belong here. I do *not* say "See, my arm is going up!' when I raise it.[10]

The therapeutic aim that the philosophy of psychology carries out consists in a critical attitude regarding introspective knowledge or intuitions, and a correspondent concentration on the multiple language games with which the inner life is expressed and described. In order to exercise a therapeutic criticism of psychology, philosophy of psychology (in the wittgensteinian sense) has to identify, we would say, the particular material of psychology that interests to that philosophy of psychology. *Philosophical Investigations* 571 gives us important information about the status of psychology subjected to the therapeutic analysis:

---

9 Wittgenstein PI, § 618.
10 Wittgenstein PI, § 627.

> Misleading parallel: psychology treats of processes in the psychical sphere, as does physics in the physical. Seeing, hearing, thinking, feeling, willing are not the subject of psychology *in the same sense* as that in which the movement of bodies, the phenomena of electricity, etc., are the subject of physics. You can see this from the fact the physicist sees, hears, thinks about, and inform us of these phenomena, and the psychologist observes the *external reactions* (the behaviour) (*Äusserungen, Benehmen*) of the subject[11].

So philosophy of psychology emphasizes the importance of the notion of expression (*Äusserung, Ausdruck*) and its significant difference in relation to the notion of description. This is a fundamental element in the wittgensteinian philosophy of psychology and really critical regarding the correspondent criticism of the metaphysical picture of inner life. It is also critical for the understanding of intention and intentionality. On the one hand Wittgenstein doesn't refuse the role of first person in the intentional action, since all intentional action is first of all expressive and expression is necessarily adjudicated to the first person perspective. On the other hand the subject that expresses intention to do p has not a so-called epistemic privilege about his mental acts. Expressive privilege is not a sort of epistemic authority, which is denied to other people. It is a nonsensical language-game to state that "I know that I intend to do p" as well as "I know that I believe p", but it makes perfect sense to say that "I know he intends that p" or "I know he believes that p". Famously Wittgenstein asserts that

> I can know what someone else is thinking, not what I am thinking. It is correct to say that "I know what you are thinking", and wrong to say "I know what I am thinking". (A whole cloud of philosophy condensed into a drop of grammar[12].

And the primacy of expression as a first person experience over the description is also pointed out in these terms: "My own relation to my words is wholly different from other people's... If I listened to the words of my mouth, I might say that someone else was speaking out of my mouth". So an expression of intention or of will have, say, their roots in the peculiar exteriorisation of a first person. Expressing an intention is not to make a report or a description and the

---

11 Another reference to the status of psychology shows how Wittgenstein distances his view from a traditional behaviouristic stance. Is it not in some measure psychology a study of behaviour as well as the physics is the study of the behaviour, say of a point of light on a screen? "Then psychology treats of behaviour, not of mind? What do psychologists record? – What do they observe? Isn't it the behaviour of human beings, in particular their utterances? But *these* are not about behaviour" (Wittgenstein PI, II, v, 179).

12 Wittgenstein PI, II, xi, 222.

metaphysical picture supposes that it is the same thing expressing an intention and to make a description. Anyway there is certainly what we call a description of mental states, namely intentional states without which it would not be possible to speak of an inner life. As well it is possible to represent intentions about the future, to make predictions of my behaviour and other person's behaviour.

3 The Therapeutic of Philosophy of Psychology

Now it seems to me that the therapeutic of philosophy of psychology contains as a relevant element the clarification of the so-called inner life, the analysis of the use of intentional statements in two axes of time, past and future. In other words, *Philosophical Investigations* carries out its therapeutic work of intention expressions, mainly through the analysis of two language-games: retrospective intentions and intentional predictions. Description of past intentional acts, as well predictions of future intentional acts correspond to different kinds of uses of language essential to our human form of life.

Let's consider the first kind of language-games of past intentional acts. Sections from 633 on until the last section of *Philosophical Investigations* present usual situations in which one describes his own *past intentions*. The subject not yet submitted to therapy speaks of his own experience of intention in the past as he supposes that he can capture in the whole aggregate of his inner life an essential and detachable point. In these last sections it is striking the focus on the link between *intention and memory*. Hacker notes that

> ...the sequence of thirty-two remarks concerning intending on the *Investigations* is focused primary on the theme of remembering what one was going to do or to say[13].

If there is not inner life without intentional actions, and if it is essential to intentionality its temporal dimension, then the status of past intentional acts are most important to analyse. In the simple conversation, for example, I wish to report a past intention (a more or less recent one) and describe it to my interlocutor. The interlocutor asks for the status of the past intention: "You were interrupted a while ago; do you still know what you were going to say?" – If I answer that, yes I'm sure I had that very thought or intention, what I try to do at the end is to *remember* a point that I experienced, a point that now I try to capture among various details that appear to be part of a scene that I represent in retro-

13 Hacker 1996, 625.

spective. Section 635 exposes this interesting movement (trying to capture my past intention):

> 'I was going to say ...' – You remember various details. But not even all of them together shew your intention. It is as if a snapshot of a scene had been taken, but only a few scattered details of it were to be seen: here a hand, there a bit of a face, or a hat – the rest is dark. And now it is as if we knew quite certainly what the whole picture represented. As if I could read darkness[14].

This beautiful written remark places the past intentional point as a trace in the whole of a snapshot, whose pieces are scattered by the work of memory. And yet the language-game also works in a way that "it is as if we knew quite certainly what the whole picture represented". Philosophy of psychology interprets the work of the language-game and the correspondent therapeutic movement shows, in this case, how language creates a kind of pure intentional point where there is only scattered pieces emerging from the darkness of memory. *Past intentional experiences are past remembrances.* What happens is that remembering a past intentional act places it in this snapshot like any other piece scattered through that picture: you have not a pure intention image that you can separate or detach from the snapshot. Therefore Wittgenstein points out what his philosophy (namely his philosophy of psychology) constantly underlines: we need become conscious of the difference of language-games because the clothing "of our language makes everything alike"[15]. So "I intended do p" seems to be about the same thing (to have the same content) as to say "I remember that I intended to do p". But it is also very different since remembering an intention places us at the centre of the complicated work of memory. Each time we want to remember an intentional experience, or the inner experience of intention, "it is (in the words of Wittgenstein) as if one had altered the adjustment of a microscope. One did not see before what is now in focus"[16].

Also the case of expression of future intentions is remarkable in order to understand the intentional stance from the therapeutic perspective of the philosophy of psychology. We are speaking of predictions of intentional acts, which are language-games very similar to predictions of voluntary acts, to which the *Philosophical Investigations* dedicates some sections from 611 on (and note *a propos* the following remark referring to the prediction of voluntary acts, namely our voluntary acts):

---

14 Wittgenstein PI, § 635.
15 Wittgenstein PI, II, xi, 224.
16 Wittgenstein PI, § 645.

> When people talk about the possibility of foreknowledge of the future they always forget the fact of prediction of one's own voluntary movements[17].

This kind of predictions is very important for all our inner life. A singular characteristic of these predictions is twofold, in first place it doesn't express any epistemic authority of the subject, since other persons can also correctly anticipate the predicted behaviour, in the second place, these kinds of predictions are fallible, not only from the first as well as from the third person perspective. "It is possible to imagine a guessing of intentions like the guessing of thoughts, but also a guessing of what someone is actually *going to do*. To say 'He alone can know what he intends' is nonsense: 'to say 'He alone can know what he will do', wrong. For prediction contained in my expression of intention (for example 'When it strikes five I am going home') need not come true and someone else may know what will really happen" and Wittgenstein pursues:

> Two points, however, are important: one, that in many cases, someone else cannot predict my actions, whereas I foresee them in my intentions; the other, that my prediction (in my expression of intention) has not the same foundation as his prediction of what I shall do, and the conclusions to be drawn from these predictions are quite different[18].

In other terms, only I can express my intentions about my behaviour in the future, but that doesn't mean that other people cannot predict that behaviour and that only I can make that prediction. Once more I have not epistemic privilege of first person, although here it would plausible to find a contradiction with the fact recognized by Wittgenstein that many times other people cannot predict my actions, whereas I foresee them. Is there not here an epistemic privilege? Well the predictions, mine and of the other person are based on different foundations, there is no doubt. But both refer to events that cannot occur.

Regarding fallibility both predictions (from my first person perspective, as well as from the third point of view) have the same exposure to fallibility. It is just the case that a "prediction contained in my expression of intention (for example 'When it strikes five I am going home') need not come true". It is also interesting to remark the following characteristic of these kinds of predictions. They also depend on memory as well as the intentions referring to the past were an exercise of memory. In fact when I say that in five minutes I'll intend to raise may arm that implies that at that very future moment I have to remember what was predicted by me five minutes before. In Anscombe view the expressions of

17 Wittgenstein PI, § 629.

18 Wittgenstein PI, II, xi, 223-4.

predictions, are an important part of intention language-games but as she notes,

> ...indeed, quite generally, the applicability of the question 'Why?' to a prediction is what marks it out as an expression of intention rather than an estimate of the future or a pure prophecy[19].

Yet to my mind she overlooks the specificity of the case of predicted intentional actions, which have always a correspondent expression in the first person and are characterized by its fallibility.

4 Some conclusions

1. Wittgenstein's philosophy of psychology must consider the fundamental difference between expression and description of experience (*Erlebnis*).
2. Philosophy of psychology has a subject–matter and this is the expression or exteriorisation of feelings, mental states, and intention. But intention is not to be confounded with any feeling or mental state.
3. In the philosophy of psychology the notion of *expression, exteriorisation* is a primary one, and it exists either as *expressed experience*, or expressed will or intentionality. This is the material that psychology studies.[20]
4. Philosophy of psychology challenges the metaphysical picture of will and intention as a *primum movens*, and thereby challenges the notion of decision.[21]
5. Philosophy of psychology considers the difference between "I raise my arm" and "the fact that my arm is raising" as a grammatical difference. This is a striking difference between propositions with the same content.
6. Philosophy of psychology challenges the so called *authority of first person* from an epistemic point of view. In this case "I know that I'm intending" is misuse of "knowing that p".
7. Philosophy of psychology carries out its therapeutic work on intentional action, through analysis of intention language-games of past intentions and intentional predictions.

---

19 Anscombe 1957, 90.
20 Wittgenstein PI, § 571.
21 Wittgenstein PI, § 620.

5 References

Anscombe, G. E. M., 1957, *Intention*, Oxford, Blackwell.
Hacker, P.M.S., 1996, *Mind and Will*, Oxford, Basil Blackwell.
Wittgenstein, L. 1980, *Remarks on the Philosophy of Psychology*, G. E. M. Anscombe, G.H. von Wright (eds.), Oxford, Blackwell.
Wittgenstein, L., 2009, *Philosophical Investigations*, 4th edn, ed. P. M. S. Hacker and J. Schulte, trans. G. E. M. Anscombe, P. M. S. Hacker and J. Schulte, Oxford, Wiley-Blackwell.

Margit GAFFAL

Intention and '*Absicht*'

Two Sides of the Same Coin

1 Introduction

If we ascribe someone an intention we presuppose that his actions are directed towards a goal. The origin of the word "intention" is derived from the Latin word "*tendere*" which means "to strain" or "tense" (a muscle) and together with the prefix "in" the word denotes someone's "focusing on" or "being directed at an object". If someone intends to do something one is directed towards an objective. Whereas intention is a set of coordinated acts as part of an overall endeavour to achieve X, the objective is X itself.

In English the expression intention is used for a determination to act in a certain way. The related expression 'purpose'[1] denotes as well an intention or plan but additionally points to the reason for an action. German has the two expressions 'Intention' and '*Absicht*'.[2] Both are synonymous but with '*Absicht*' it is usually assumed that measures have been taken to achieve an envisaged objective. As such it was particularly used in Austrian officialese of the state administration in the late 19th and at the beginning of the 20th century. Whereas "intention" points to a person's orientation or tendency towards an aim, "*Absicht*" implies an elaborated plan ready to be put in practice. The word "*Zweck*" corresponds to purpose in English and refers to aspects, such as the function, benefit or application of an objective. The following table gives an overview of the expressions in the semantic field of intention:

1 The word consists of the word root "pose" originally meaning an act of posing the body and was used in English from the early 15th century in the sense of "assuming a certain attitude".[1] It is combined with the prefix "pur" derived from Latin pro- meaning 'forth'.

2 The word "*Absicht*" is derived from the verb "*sehen*" (Middle High German: "*sehan*") which developed from the Latin word "*sequi*" which means "to follow" something (with eyes) in the context of hunting. In the 17th century the word "*absehen*" was used in the sense of "to measure", or "to estimate" and also meant "destination, goal". In the 18th century it acquired the connotation of "endeavour" and "volition".

https://doi.org/10.1515/9783110560282-109

Language			Words having a family resemblance		
German	Intention	*Absicht*	*Zweck*: 1. *Funktion* 2. *Nutzen* 3. *Verwendung*	*Ziel*	*Motiv*, *Ursache* *Grund*
English	intention	purpose plan	purpose 1. function, 2. benefit, utility, 3. application	aim, purpose, objective, target	motive cause reason, ground

The translations suggested here are based on a study of the different contexts in which these words are used. Ludwig Wittgenstein addressed some problems regarding the problem of intention. In his writings in German he used mainly the word '*Absicht*' to analyze the mode in which intentional action can be expressed, described and fulfilled. A revision of the *Philosophische Untersuchungen* shows that Wittgenstein preferred "*Absicht*" (60 times to be precise) to investigate the question of when we act on purpose, and used "Intention" only on three occasions. One reason may be that with "*Absicht*" there is a verb related "*beabsichtigen*" whereas there is no such verb in German with intention. No distinction is made between "Intention" and "*Absicht*" in the English translations. Anscombe's translation contains sometimes the overall concept of 'intentionality'[3] though it was actually never used by Wittgenstein. This article traces Wittgenstein's most relevant questions of what it means to have an intention and provides a summary of his arguments. The following section shows how the meaning of the expressions intention and *Absicht* developed over time.[4]

2 Etymology of Intention and '*Absicht*'

Words have a long history in the course of which original meaning underwent a modification or even a complete change. A diachronic study allows us to ascer-

3 Wittgenstein 2001, § 35.

4 The word "*Intentionalität*" was not used at all in the PU. The English translations do not distinguish between "Intention" and "*Absicht*". On some occasions Anscombe used the English word "intentionality" to translate "*Absicht*" (for instance in Wittgenstein 2001, § 35) but that does not exactly correspond to the original text in German.

tain the meaning of words in different historic contexts. In the case of "*Absicht*" we come across the middle high German word "*sehan*" that derived from the Latin word "*sequi*" meaning "to follow" something (with eyes), which was mainly used in the context of hunting.

In Roman jurisprudence, Cicero distinguished two categories of acts, intentional acts (*voluntaria*) and acts that result from lack of knowledge (*ignorantia*). In the second category human acts may either be caused by chance (*fortuna*) or by necessity (*necessitas*). Intentional acts are linked to knowledge in the sense that only if someone has knowledge he is capable of intentional acts. For instance, Hadrian and Cicero recommended judges when rendering verdicts for civil wrongs that they should mainly focus on the defendant's intention rather than on the consequences of an action.[5] The reason is that an intended aim may diverge substantially from the actual outcome of an action.

In the 17th century the word "*absehen*" was used in the sense of "to measure" and "to estimate" and was mainly applied to the development of manual skills and handcraft. Simultaneously a second meaning evolved to denote "destination" or "goal" in a more abstract sense. In the 18th century it was further refined and acquired the connotation of "endeavour" and "volition".

Grimms '*Wörterbuch*' contains an entry on "*Absicht*" and all its derivations: the verb "*absichten*", the synonymous adjectives "*absichtsreich*" and "*absichtsvoll*", its antonym "*absichtslos*", the adverb "*absichtlich*", and the noun "*Absichtlichkeit*". The entries contain references to various authors, such as G.E. Lessing, F. Schiller,[6] J. W. von Goethe,[7] Wieland, J. G. Fichte and I. H. Voss. In his translation of Homers '*Illias*' from Greek to German, Voss describes Ulysses' return to Ithaca. When he asks a shepherd without being recognized what had happened to Penelope and his son, the shepherd describes the intruders as

---

5 "Hadrian verfügt "malef, maleficiis voluntas spectatur, non exitus" (bei Verbrechen soll man Willen beachten, nicht den Tatausgang) (...) und Cicero: animum, non eventum considerent" (sie sollen die Absicht, nicht das Tatergebnis berücksichtigen)." Ueding 1994, 129.

6 See Goethe's *Torquato Tasso*, in which Tasso speaks to the princess: "Wenn sie auch die Absicht hat, den Freunden wohlzutun, so fühlt man Absicht und man ist verstimmt." Goethe 1977, 99.

7 During his Italian Journey, J. W. von Goethe wrote a German version of the ancient Greek tragedy '*Iphigenia in Tauris*', where Iphigenia complains to Arkas about the King's lack of tact and sensitivity in his courtship towards her. However, Arkas tries to convince Iphigenia not to reject the King justifying his misbehaviour by reminding her of the King's misfortune of having lost his son and therefore he was unable of evoking trust in a conversation.

"intentionally hiding the truth" when entering Ulysses' house and intending to win Penelope's favour.[8]

In the Age of Enlightenment, G. E. Lessing used '*Absicht*' to denote *a higher-order quality* to distinguish the determined individual from simple mankind.[9] At that time the expression "in Absicht des..." also had the adverbial function of "concerning, regarding' or 'with reference to. For instance, in his '*Reden an die deutsche Nation*', J. G. Fichte pointed to the importance of a proper form and content when education the younger generation.[10] In first coming-of-age novel (Bildungsroman) '*Die Geschichte des Agathon*' the German writer C. M. Wieland used "*Absicht*" to describe the sophists' intention to rule about other people's minds and thereby abusing them for their own ends.[11] What he calls higher intentions refers to the sophists' skill to assume any form at will to inspire trust in others aimed at subsequently manipulating them.[12]

In the late 18th and early 19th century the word "*Absicht*" formed part of Austrian officialise and can be found in documents of the state administration. It is used in official reports of public planning, spending or restructuring such as in

---

8 When Ulysses wants to sound out the shepherd he received the shepherd's following response: "Greis, kein irrender Mann, der jenen verkündend herkommt, wird leicht Glauben gewinnen bei seinem weib' und dem sohne. Nur zu oft, dass nach pflegeverlangende länderdurchwandrer Teuschen, und absichtsvoll im Gespräch ausweichen der wahrheit. Denn wer, die welt durchirrend, in Ithaka's fluren gelanget, meiner gebieterin nahet er gleich, und schwazet erdichtung: Doch sie empfängt und bewirtet ihn wohl, und forschet nach allem." (Voss 1806, 125).

9 "Mit Absicht handeln ist das, was den Menschen über geringere Geschöpfe erhebt; mit Absicht dichten, mit Absicht nachahmen, ist das, was das Genie von den kleinen Künstlern unterscheidet, die nur dichten, um zu dichten, die nur nachahmen, um nachzuahmen...". (Lessing 1996, IV, 388 f.).

10 "Diese Erziehung erscheint nun nicht mehr, (...) sondern den ganzen Menschen durchaus und vollständig zum ganzen Menschen zu bilden. Hierzu gehören zwei Hauptstücke: zuerst in Absicht der Form, dass der wirkliche lebendige Mensch, bis in die Wurzel seines Lebens hinein, (...) gebildet werde; sodann in Absicht des Inhalts, dass alle nothwendigen Bestandtheile des Menschen ohne Ausnahme und gleichmässig ausgebildet werden." (Fichte 1846, 301).

11 "Die Kunst über die Einbildungskraft der Menschen zu herrschen, die geheimen, ihnen selbst verborgenen Triebfedern ihrer Bewegungen nach unserm Gefallen zu lenken und sie zu Werzeugen unsrer Absichten zu machen, in dem wir sie in der Meinung erhalten, dass wie es von den ihrigen sind, ist also ohne Zweifel diejenige, die ihrem besitzer am nützlichsten ist, und dies ist die Kunst welche die Sophisten lehren und ausüben." (Wieland 1818, 170).

12 "Diese Überredungskraft setzt die Geschicklichkeit voraus, jede Gestalt anzunehmen, wodurch wir demjenigen gefällig werden können, auf den wir Absichten haben." (Wieland 1818, 178).

announcements of envisaged programs ready to be implemented.[13] The distinction between intention and "*Absicht*" becomes obvious: whereas intention points to an individual's disposition to produce a certain result, "*Absicht*" involves a detailed description of how the objective can be achieved whereby the individual disposition is secondary. In the case of intention the focus is on the individual's readiness directed towards an aim whereas "*Absicht*" includes the operational steps and specific goals that are aimed to produce a certain result.

In 1874 F. Brentano introduced 'Intention' and used it systematically in his treatise entitled '*Psychology from an Empirical Standpoint*' (*Psychologie vom empirischen Standpunkt*). He applied the expression "*intentionale Inexistenz*" as a criteria to distinguish between physical and psychological acts. As such, he characterized psychic phenomena as being related to what he called the intentional inexistence of objects. What an object stands for is not a material substance but content or what he calls immanent objectivity. Every psychological phenomenon contains something as an object in itself, the same as in a judgment something is recognized, in love something is loved or else something is hated in hatred. He described this in the following quotation:

> Every mental phenomenon is characterized by what the Scholastics of the Middle Ages called the intentional (or mental) inexistence of an object, and what we might call, though not wholly unambiguously, reference to a content, a distinction upon an object (which is not to be understood here as meaning a thing), or immanent objectivity. Every mental phenomenon includes something as object within itself, although they do not all do so in the same way. In presentation something is presented, in judgement something is affirmed or denied, in love loved, in hate hated, in desire desired and so on.
> This intentional in-existence is characteristic exclusively of mental phenomena. No physical phenomenon exhibits anything like it. We could, therefore, define mental phenomena by saying that they are those phenomena which contain an object intentionally within themselves.[14]

---

13 See: "Kaiserliche Verordnung vom 5. Dezember 1848 an den provisorischen Minister des öffentlichen Unterrichts, wodurch die Absicht erklärt wird, die Aufhebung de k.k. Theresianischen Ritterakademie auszusprechen. Da Ich die Absicht habe, die von Ihnen in Antrag gebrachte Aufhebung der Theresianischen Ritterakademie auszusprechen, so erwarte ich die Mir zugesagten Vorlagen in Absicht auf die zukünftige Bestimmung des Gebäudes für den öffentlichen Unterricht, auf die Verwendung der Fonde und Stiftungen, sowie auf die Behandlung auf die in der Akademie befindlichen Zöglinge und angestellten Professoren, Beamten und Diener. Franz Joseph m.p." (22 July 2016: <http://alex.onb.ac.at/cgi-content/alex?aid=rgb&datum=18490005&seite=00000011>).

14 Brentano 1995, 88 f. Brentano 2008, 106.

To sum up: the etymology of intention shows that its emergence was related to the development of skills used to make a living. It involved knowledge that contributed to the development of society. Since Roman times the notion of intention has been linked to the legal concept of responsibility. The difference between intention and '*Absicht*' is that the former seems to have a more abstract meaning whereas the latter is linked to practical measures on the way to an intended result. In the following section the different modes by which intention is communicated will be examined.

3 How is Intention Expressed?

Any meaningful action is thought of, planned and carried out by using language. If we assume that someone acts intentionally and not just accidentally he is supposed to achieve his aim. Any intention is linguistically created and its related actions go hand in hand with language use. But how is intention expressed in language? How does a person whose actions are directed towards a purpose put his intention in words? One is inclined to perceive action as a temporary activity situated between two poles, from the original thought or plan over a set of acts that allow for the achievement of a goal. The steps taken in between these two poles may be viewed as acts that require a continuous coordination and adaptation to the circumstances.

An individual intention may interfere with other people's intentions and thereby produce frictions. Once an agent discloses his intention to others within a social framework the others may put obstacles to that aim. Therefore people tend to hide their intentions. We shall exemplify this in the following case: let us assume person A intends an action X and B observes A's conduct. As a consequence B detects A's intention and discloses it. A is aware that B has recognized his intention. As a counter-reaction to B's prediction A refrains from the intended action and leaves X undone. As B never receives a confirmation of his having correctly guessed A's intention, he has two possibilities: (1) B believes to have recognized A's intention though the proof is missing. B interprets that A has refrained from the action and impeded its proof. B will probably search for further evidence of A's intention. (2) B concludes that he has incorrectly ascribed an intention to A. As a consequence B may doubt his powers of observation and might be more cautious next time. Wittgenstein describes a similar mechanism in the following quote in which he reflects about a person's ability to detect another's intentional thoughts:

> It is possible to imagine a guessing of intentions similar to the guessing of thoughts, but also a guessing of what someone is actually *going to do.*
> To say "Only he can know what he intends" is nonsense; to say "Only he can know what he will do", wrong. For the prediction contained in my expression of intention (for example, "As soon as it strikes five, I'm going home") need not come true, and someone else may know what will really happen.[15]

The one who has an intention seems to have an initial advantage because he perceives the events from the first-person-perspective and oversees both, his plan to achieve the goal and other people's possible (counter-)reactions to it. The person aiming to detect another's intention observes the other's conduct and contrasts it with the language used along with actions. An agent intending something is supposed to follow a certain procedure or rule according to which he thinks he will reach his aim. Still, he is free to change his conduct at any time to dissimulate or hide his intention. To ascribe an intention to another person requires that the person making an ascription detects indications suggesting said ascription. Such indications are interpreted in line with the observing person's background knowledge. The question is whether agent and observer coincide in their knowledge and language use. Both, agent and observer may concur in a certain intention, but their considerations are based on different starting points. Wittgenstein described this in the following quote:

> Two points, however, are important: one, that in many cases someone else cannot predict my actions, whereas I foresee them in my intention; the other, that my prediction (in my expression of intention) does not rest on the same foundation as his prediction of my action, and that the conclusions to be drawn from these predictions are quite different.[16]

Someone's disclosure of intention may create problems. But even if one hides an intention keeping it secret it is most likely that one's conduct reveals the underlying intention. This is because the plan to reach an aim is thought of in language and therefore follows a rule or at least a certain pattern that one has acquired beforehand. In line with Wittgenstein's view that a form of life is expressed by language games,[17] we assume that a person's mode of expression reveals information on several aspects, such as the vocabulary that is used to address an issue, the links created among linguistic structures and the order of a sequence of actions, etc. The grammar of the person's enunciations shows us how this person thinks, plans and approaches things. Language is a social

---

15 Wittgenstein 2001, *Philosophy of Psychology - A Fragment* xi, § 328.
16 Wittgenstein 2001, *Philosophy of Psychology - A Fragment* xi, § 329.
17 Padilla Gálvez, Gaffal 2011 and Padilla Gálvez 2016.

technique that we have acquired to interact with our environment and we cannot just unlearn or forget it. We have not learned the meaning of words in a private setting but in the public sphere and it is a tool that we use to capture, describe and act upon the world. Wittgenstein mentioned this on several occasions, two of which are quoted here:

> *Essence* is expressed in grammar.[18]
> (...) man kann nicht die Sprache (Technik), die man beherrscht, nicht beherrschen, willkürlich verlernen.[19]

The language a person uses marks an important point of reference to comprehend the content of his enunciations. We shall analyze the mode in which intention is communicated in different language games. The context of language games is examined in order to reconstruct the implicit information underlying these enunciations. In German an intention is usually communicated by one of the following enunciations, each of which has a different connotation:

> (1) Ich will X unter Bedingungen Y tun. (*Wille*)
> (2) Ich möchte X unter Bedingungen Y tun. (*Wunsch*)
> (3) Ich werde / plane / habe vor / beabsichtige X unter Bedingungen Y zutun. (*Absicht, Vorhaben, Plan*)

Sentence (1) expresses a will, whereas sentence (2) points only to a wish or desire and sentence (3) states a more concrete purpose, endeavour or plan. The English translation requires some explanatory remarks since '*wollen*' does not fully correspond to the English verb "will", as this has lost its significance to express volition and is mainly used to form future tense.[20] Instead of "will" other verbs are used, such as "want, wish, aim to, intend to or mean". A distinction is made between intention, wish and plan:

> I want / intend / plan to / aim to do X in circumstances Y. (*intention, will, volition*)
> I wish / would like to do X in circumstances Y. (*wish, desire*)
> I will / shall / am going to do X in circumstances Y. (*plan in near future*)

18 Wittgenstein 2001, § 371.

19 Wittgenstein 2000, Item 37, 63b.

20 Yet, it is still in use to express a firm determination or someone's being desperate to do something, in "He wills to be good" (*Er hat den festen Willen, gut zu sein*). See for instance the enunciations: "If you really will a thing, that thing will be" (*Wenn man etwas wirklich fest will, dann wird das auch zustande kommen*).

One expresses a will or intention by saying (1) "I want / intend to / mean to do X in circumstances Y". The same as in German the speaker does not establish a reference to the environment but simply reports a personal will or intention. The same is true for (2) in which one is asked about one's wish or desire. Neither of the enunciations in (1) and (2) can be verified by the recipient nor can the agent be held responsible for his statement.

If we consider enunciation (3) "I will / shall / am going to do X in circumstances Y" it is usually assumed that one has firmly decided to do something. It suggests an action to happen in the near future. The action is announced, yet its content is still unaccomplished at the time of speaking and can only be fulfilled in the future. The recipient of such message may, however, control its completion at a future point of time.

If a person's intention is expressed in the form of (1) "*Ich will X unter Bedingungen Y tun*", then the focus lies on the subject rather than on the intended object X. We are only informed about a person's possible good, strong or free will to carry out or reach something under certain conditions. Expressing intention by "*wollen*" directs us towards the first person-perspective, as an analysis of the word "*Wille*" reveals. It has originally derived from Latin "*velle*" and was converted into "*willio*" (Old High German) and over time became "wille" (Middle High German).[21] The early use of "*wollen*" was to express a person's consent or agreement to do something.[22] In the 19th century "*Wille*" was combined with "*Akt*" (Latin "*acta*", result of negotiations) to form the compound "*Willensakt*" which denotes an officially documented act of will.[23] If intention is expressed as an act of volition we are focused on the person's supposed motives rather than on the object or purpose. For Wittgenstein the act of volition and its fulfillment are not separated but form a unity. A person's action is always a reflection of volition. To want to do something without doing it makes no sense, it contradicts the content of the word "want" he says:

> Verhält es sich also so daß ich meine Handlungen mit meinem Willen nur begleite? Aber wie kann ich dann voraussagen – und das kann ich doch in einem gewissen Sinne – daß ich in 5 Minuten meinen Arm heben werde? Daß ich dies wollen werde? Es ist klar: Es ist unmöglich zu wollen ohne nicht schon den Willensakt auszuführen. Der Willensakt ist

---

21 From this word-root resulted the two verbs "*wollen*" and "*wählen*" (choose).

22 Middle High German "*willec*", nowadays "*willig* sein".

23 Another derivation is "*Willkür*" which means arbitrariness and implies that the criteria for an action are hidden or unclear, even without regard to principles or laws.

> nicht die Ursache der Handlung sondern die Handlung selbst. Man kann nicht wollen ohne zu tun.[24]

According to his view a person's volition finds its expression in the execution of a desired action. It would be meaningless for someone to express his intention to do something without doing anything. And he also says this:

> The fact that I will an action consists in my performing the action, not in my doing something else which causes the action.
> When I move something I move.
> When I perform an action I am in action.[25]

To sum up: the concept of volition seems far too limited to cover the whole process involved in intention, from a first thought over the steps to reach sub-goals to what is finally expected to happen. The focus is only on the person but neglects the intended aim. We use a particular vocabulary and specific grammatical structures to express intention. It is characteristic of any intentional expression that its content is open and can only be fulfilled in the future. This is the main difficulty when ascribing an intention to others. Whereas with "*Absicht*" we keep an eye on the envisaged aim with intention we tend to focus more on agent's motives for his or her aim or goal.

4 Ascribing Intention to Somebody

The problem with intention lies in the fact that it takes place inside a person's mind. Therefore one can only speculate about a person's intention and search for empirical evidence to underpin one's assumption. It is a problem of demonstrability the same as we encounter in the case of a person's feelings. In the following quote Wittgenstein compares intention with emotions such as anger:

> Die Intention ist nun aber so genau derselben Art wie — z.B. — der Ärger. Und da scheint es irgendwie als würde man die Intention von außen betrachtet nie als Intention erkennen; (...) Das ist natürlich wieder das vorige Problem denn der Witz ist daß man es dem Gedanken (als selbständige Tatsache betrachtet) ansehen muß daß er der Gedanke ist,

---

24 Wittgenstein 2000, Item 103, 73r.

25 "Daß ich einen Vorgang will besteht darin daß ich den Vorgang mache nicht darin daß ich etwas anders tue was den Vorgang verursacht.
Wenn ich etwas bewege so bewege ich mich.
Wenn ich einen Vorgang mache so gehe ich vor." Wittgenstein 1979, 88.

> daß das und das der Fall ist. Kann man es ihm nicht ansehen (...) dann hat er kein logisches Interesse oder vielmehr, dann gibt es keine Logik.[26]

The question is whether one can ever know a person's intention. And if so, which evidence would be reliable to give account for an intention. The answer lies in language use because it reveals a person's state of mind. As language game correspond to a form of life and intention is embedded in this form of life. The use of language games requires that we have learned how the language is to be used. We prefer certain language games depending on which language games we have learned and practiced. It is part of this learning process that the speaker acquires also certain customs and habits. Thus, each speaker has a linguistic profile, which allows for an unlimited possibility of variations. However, a person's linguistic profile predominates over these unlimited possibilities and emerges as a typical pattern. The concept of intention was introduced and acquired in the same way as language games were learned and practiced.[27] The manner in which a speaker expresses his intentions linguistically, such as for instance, the degree of accuracy or obligation, whether in the form of a future statement or a wish, reveal how the speaker has learned to act. Wittgenstein describes this in the following quotation:

> But didn't I already intend the whole construction of the sentence (for example) at its beginning? So surely it already existed in my mind before I uttered it out loud! - If it was in my mind, still it would not normally be there in some different word order. But here again, we are forming a misleading picture of 'intending': that is, of the use of this word. An intention is embedded in a setting, in human customs and institutions. If the technique of the game of chess did not exist, I could not intend to play a game of chess. To the extent that I do intend the construction of an English sentence in advance, that is made possible by the fact that I can speak English.[28]

In his aim to detect characteristic features of intention in human behaviour, Wittgenstein applied the method of inversion of an argument (*argumentum e contrario*), which we encounter on various occasions in his writing. If he has so far considered the existence of an internal relation necessary, he now opts for the opposite and looks what happens when negating the whole concept. The underlying thought is this: in principle, intention is supposed to be present in human behavior, but if it cannot be shown, it simply means that it does not exist. He puts it like this:

---

26 Wittgenstein 2000, Item 108, 187.

27 Padilla Gálvez, Gaffal 2011, 9 ff.

28 Wittgenstein 2001, § 337 (*Philosophische Untersuchungen, Spätfassung* (TS 227), 900).

> Die Intention muß natürlich auch ein Phänomen sein. D.h. Wenn man alle Phänomene in Betracht zieht und die Intention würde sich in ihnen nicht zeigen so wäre sie auch nicht da.[29]

To illustrate this he uses a curious example from applied mechanics and describes a machine which is equipped with a brake arm (*Bremshebel*) that is 'intended to stop the machine'. Unfortunately, this brake arm does not work properly and functions only occasionally and at irregular intervals. Due to this irregular functioning, a layman is unable to detect the real function of the brake arm, namely that it was designed to stop the machine. A specialist who knows about the brake arm's function of operation and now monitors the machine becomes angry and shows an annoyed reaction each time the brake arm fails. An uninvolved person who observes the scenario and who does not know the concept of intention now perceives the following procedure: sometimes he sees the machine stopping as a consequence of the brake arm's functioning and sometimes he sees an angry man.[30]

Wittgenstein approached the problem of intention by applying the principle of causality as in physics or chemistry, in which an action provokes a reaction (*actio-reactio*). On the contrary, what we are confronted with here is finality and no causality. The principle of cause-effect is applied in the natural sciences to explain the mutual relationship between objects and processes. An objective cause provokes another event. But human behaviour cannot be explained by causality. In the following quotation, Wittgenstein recognizes this shortcoming. Therefore the relevant question is how thoughts become manifest. In other words, how can one ever know what the other thinks? He says this:

> Und hier haben wir natürlich das alte Problem daß nämlich der Gedanke daß das und das der Fall ist nicht voraussetzt daß es der Fall ist. Daß aber andererseits doch etwas von der Tatsache für den Gedanken selbst Voraussetzung sein muß. "Ich kann nicht denken daß etwas rot ist wenn rot gar nicht existiert". Die Antwort darauf ist daß die Gedanken in demselben Raum sein müssen wie das Zweifelhafte wenn auch an einer anderen Stelle;

---

29 Wittgenstein 2000, Item 108, 185.

30 "Wenn nun eine bestimmte Person immer dann wenn der Hebel nicht als Bremshebel wirkt, ärgerlich wird -so wäre damit auch nicht das gezeigt was ich zeigen will. Ja man könnte dann sagen daß der Hebel einmal die Bremse, einmal den Ärger betätigt. – Wie drückt es sich nämlich aus, daß die Person darüber ärgerlich wird, daß der Hebel die Bremse nicht betätigt hat? (Dieses über etwas ärgerlich sein ist nämlich scheinbar von ganz derselben Art wie etwas fürchten, etwas wünschen etwas erwarten etc.)" Wittgenstein 2000, Item 108, 185.

> daß die gegenwärtige Realität auf die der Gedankenmaßstab aufgestellt wird den Sinn – nicht verbürgt – sondern ausmacht.[31]

He tries to solve the problem by investigating emotions, such as anger, fear or wishful thinking. The reason for feelings does not need to be a fact but there must be something on which a feeling is based. G. Frege reminded us that the meaning of a sentence arises from its underlying thought.[32] If a person is afraid or annoyed, one can reasonably assume that he is aware of what he is afraid of or worried about. To want something without knowing what one wants or expecting something without knowing what one expects makes no sense. The meaningful connections we make are based on real conditions. He says:

> Der Sinn kann ebensowenig erst verbürgt werden müssen wie es nachträglich bewiesen werden kann daß π nicht rational ist; denn ohne Sinn kein Gedanke. – Darin und nur darin besteht auch die (prästabilierte) Harmonie zwischen Welt und Gedanke.[33]

As a conclusion, the question of how to detect a person's intention cannot be solved by applying the principle of cause and effect. Instead of that an action can be viewed as a product of thoughts and as accompanied by language. The meaning of words is objective in the sense that each word denotes the content that is established within a speech community. Individual language games reveal the particular vocabulary and grammatical structure that a speaker uses to convey a meaningful content. Intention is embedded in these language games[34] and they gives us a hint on what a person is up to.

5 When Is an Intention Fulfilled?

Wittgenstein asked how a person having an intention in mind could recognize that the intended aim has occurred. How can one match an intention to the intended result? He suggested a possible relation between the description of the intended event on the one hand and the result on the other:

---

31 Wittgenstein 2000, Item 108, 186.

32 Frege 1892, 39 f.

33 Wittgenstein 2000, Item 108, 186.

34 Padilla Gálvez, Gaffal 2011, 7 ff.

> We use words to denote elements in reality. Our statements arise on the basis of ideas we put into words.[35]

While Wittgenstein considered a visualized representation of the intended object and its occurrence as an external relation, Russell, Ogden and Richards suggested a special relation.[36] For Wittgenstein, the mere thought and the fact are relevant criteria that help verify the coincidence. However, Russell mentioned a third element that counted as verification, such as for instance, a feeling that is provoked at the moment of this coincidence. Wittgenstein described this third element in the following quotation:

> Dieses dritte Ereignis, gleichsam die Stillung des Hungers (die zwei anderen sind der Hunger und das Essen einer bestimmten Speise), dieses dritte Ereignis könnte z.B. das Auftreten eines Gefühls der Freude sein. Es ist hier ganz gleichgültig als was wir dieses dritte Ereignis beschreiben; für das Wesen der Theorie ist das ohne Bedeutung. Die Kausalität zwischen Sprache und Handlung ist eine externe Relation während wir eine interne Relation brauchen.[37]

Wittgenstein asked what such internal relation proposed by Russell would look like. He reflected on the question of how a feeling of joy could function as a confirmation of the original intention. By making this comparison he revealed an inconsistency in Russell's theory. To recognize such third element, one would need to describe beforehand how this could be recognized. But a feeling confirming the arrival of an intended aim is a rather uncertain proof, since a feeling could confirm both, the occurrence and the non-occurrence of a result. How would one ever know for sure whether the intended aim has actually come up? He says this:

> Denn welche Mittel habe ich die ursprüngliche Abmachung mit der späteren Handlung zu vergleichen? Das Vergleichen ist es was in der Russellschen Theorie nicht vorkommt. Und das Vergleichen besteht nicht darin bei der Konfrontierung der Darstellung mit dem Dargestellten ein Phänomen zu erleben das – wie gesagt – selbst von vornherein nicht beschreibbar war.[38]

---

35 Wittgenstein 2001, § 59.

36 Bertrand Russel's philosophy was based on the doctrine of external relations according to which he viewed the world as consisting of isolated entities that do not have any fixed relations to objects. Russell 2003, 87 f.

37 Wittgenstein 2000, Item 107, 289–290.

38 Wittgenstein 2000, Item 107, 291.

What Wittgenstein criticized in Russell's theory is the missing comparison between the original intention and its corresponding event. He underlined that the visualized representation of the intended aim had existed right from beginning and cannot be arbitrarily adapted to whatever event coming up later. To illustrate this he brought up an example of chess game in which a person is defeated still claiming to have won the game because a defeat was his goal. One would argue that this player has no idea of the rules of chess. According to Russell's theory, however, each game could be regarded as a win, as long as the person merely claimed that he is satisfied with the outcome.[39] This appears absurd as such attitude contradicts not only the rules of chess but also the view that to win is the goal of any game. One could argue that playing the game is the goal, but then it would not make sense to claim to have won the game. Wittgenstein argued that the expectation of an intended event must be in harmony with its actual occurrence and cannot be the opposite. Moreover, a feeling is an inadequate element to decide whether an expectation is fulfilled. He explains this in the following quotation:

> Die Erfüllung der Erwartung besteht nicht darin daß ein Drittes geschieht das man außer als eben "die Erfüllung der Erwartung" auch noch anders beschreiben könnte, also z.B. als ein Gefühl der Befriedigung oder der Freude oder wie immer. Denn die Erwartung daß p der Fall sein wird muß das gleiche sein wie die Erwartung der Erfüllung dieser Erwartung, dagegen wäre, wenn ich unrecht habe, die Erwartung daß p eintreffen wird verschieden von der Erwartung daß die Erfüllung dieser Erwartung eintreffen wird.[40]

If one's intention finds it expression in the expectation of an event then one must have at least an idea of what fulfillment of this expectation looks like. Wittgenstein questioned a subjective feeling as proof of the occurrence of the expected insisted on an objective component as a confirmation. There must be such a connection otherwise we would expect nonsense:

> Wäre der Akt der Erwartung nicht mit der Realität verknüpft so könnt man einen Unsinn erwarten.[41]

He objected to the picture-theory that a visualized representation seems ineffective as it can be diffuse and may be interpreted in various ways. He says this:

---

39 Wittgenstein 2000, Item 107, 294.
40 Wittgenstein 2000, Item 107, 293.
41 Wittgenstein 2000, Item 107, 265.

> Wie ist das Bild gemeint? Die Intention liegt nie im Bild selbst denn wie immer das Bild beschaffen ist, immer kann es auf verschiedene Weise gemeint sein. Das sagt aber nicht daß, wie das Bild gemeint ist, sich erst zeigen wird wenn eine bestimmte Reaktion eingetreten sein wird, denn die Intention drückt sich schon jetzt darin aus wie ich das Bild jetzt mit der Wirklichkeit vergleiche.[42]

Since a visualized representation does not provide clear indications, he rejects the picture-theory and concentrates on language as a link between the intention and its realization. When verbalizing one's intention and adapting one's goals to the existing circumstances one determines how it is meant. As we have learned the language in the public sphere the meaning of words is not a private matter but undoubtedly a matter of convention and agreement within a language community. When someone articulates one's intention one sets the rules for its comprehension.

How is intention related to a person's expectation of an objective? We are only interested in those expectations that can be fulfilled, and leave aside unfulfillable expectations.[43] A person plans an action so that its execution increases the desire to reach the intended goal. In doing so, he relies on the uniformity of the space-time conditions in such a way that the planning is linked to a certain room and time. Which elements play a role? The memory is important, since one can only expect something that he knows of. Could one expect something wholly unexpected? Wittgenstein denies this when he says:

> Aber das gibt man doch zu, daß ein Unterschied ist zwischen der Erwartung und ihrer Erfüllung.[44]
> Was sich nicht hat voraussehen lassen von dem können wir eben nicht reden.[45]

One can only then speak of a thought, a wish, or expectation when these are articulated or formulated verbally. He says: "Thoughts I call the articulated process, so you can say" first that which has an articulated expression."[46] What the idea of an event has in common with its fulfillment is a definite content. This content is articulated and communicated. Here he underlines the role of

42 Wittgenstein 2000, Item 107, 293.

43 Wittgenstein says: "Ich rede hier von einer Erwartung nur als von etwas was unbedingt entweder erfüllt oder ennttäuscht werden muss, also nicht eine Erwartungs ins Balue". Wittgenstein 2000, Item 107, 258.

44 Wittgenstein 2000, Item 108, 211.

45 Wittgenstein 2000, Item 108, 212.

46 Wittgenstein says: "Gedanken nenne ich erst den artikulierten Vorgang man könnte also sagen "erst das was einen artikulierten Ausdruck hat"." Wittgenstein 2000, Item 107, 262.

language as an expression of a content that can be thought, expected and fulfilled. He explains it this way:

> Die Erwartung der Befriedigung der Erwartung daß p eintreffen wird ist die Erwartung daß p eintreffen wird. Der Gedanke an den Inhalt des Gedankens p ist der Gedanke p. Und das enthält die Wahrheit über alle die Fragen die in dieser Sache aufgeworfen werden.[47]

When can one say that an expectation is fulfilled? Let us recall the whole process: expectation is like a process of searching for an event or for a fact to occur within a space. A reference to a space is established. Wittgenstein explains this:

> Wenn ich erwarte, dass jemand zu dieser Tür hereinkommen wird, so drückt sich diese Erwartung doch ... in welchem die Tatsache erwartet wird.[48]

To conclude, the searching process leads to the expected event. An expectation can be compared to a projection in which someone projects an expected event onto a room together with a certain time frame. The expected element is first a symbol to which the person concerned approximates by means of images. "When we think of signs (symbols), we also expect and wish for signs."[49] He says this:

> Das Ereignis das die Erwartung ersetzt beantwortet sie d.h. im Ersetzen besteht die Beantwortung es kann also keine Frage geben ob das nun wirklich die Antwort ist. Eine solche Frage hieße den Sinn eines Satzes in Frage stellen. Daß ich das gefunden habe was ich früher gesucht habe kann man auf keine Weise beweisen außer dadurch daß der Fund das Suchen ersetzt. Denn auch jede Annäherung an den tatsächlichen Fund durch Bilder dessen was ich suche geschieht je nur durch das Ersetzen eines Symbols, durch ein Bild und ehe das Ersetzen stattgefunden hatte war der Ersatz nicht da und nachdem es stattgefunden hatte ist das Bild an die Stelle des Symbols getreten.[50]

A person expecting something only knows the components of a whole result but not its exact composition. Wittgenstein describes this as one's searching for a combination of elements of a complex, without knowing exactly whether this complex really exists in this form. Once the expected result is found, the symbol is replaced by the result. The expectation only lasts until the moment when the event actually occurs. If, on the contrary, an expected event X does not occur

47 Wittgenstein 2000, Item 108, 214.
48 Wittgenstein 2000, Item 107, 255.
49 Wittgenstein says: "Denn wenn wir in Zeichen denken, so erwarten und wünschen wir auch in Zeichen." Wittgenstein 2000, Item 209, 11.
50 Wittgenstein 2000, Item 107, 258.

and instead an event Y, the event Y cannot satisfy the expectation with respect to X.[51]

6 Conclusions

I have traced some of the questions that Wittgenstein raised concerning the intention of actions. According to him, a person's intention lies directly in the execution of a desired action. He combines the concept of intention with a projected image of an intended goal. The question is whether the coincidence of the picture and the intended event is an external or internal relation. Wittgenstein pleads for an external relation and emphasizes that the intention is not a posterior event, but rather linked to the picture right from the beginning. However, the picture theory proves inadequate, for a picture is too diffuse.

Instead, he concentrates on language as a link between image and reality. By describing an intended aim the agent determines the content. One can only speak of a thought, a wish, or expectation when these are articulated or verbally expressed. What the intended event and its expected fulfillment have in common is a definite content. Expectation is a process of searching for the intended result to occur within a space. Although an intention is determined by its content its fulfillment is still unknown in the sense that the agent knows its components but not exactly its form. Wittgenstein speaks of the search for a bundle of known elements of a complex. The expected is first a symbol which is finally replaced by the intended result.

7 References

Anscombe, G. E. M., 1963, *Intention*, Oxford, Basil Blackwell.

Brentano, F., 1995, *Psychology from an Empirical Standpoint*, ed. L. McAlister. London, Routledge.

Brentano, F., 2008, *Psychologie vom empirischen Standpunkt*. Frankfurt a.M., Ontos Verlag.

Fichte, J. G., 1808, Reden an die deutsche Nation, in: *Johann Gottlieb Fichte's Sämtliche Werke. Zweiter Band: Zur Politik, Moral und Philosophie der Geschichte*. Berlin, Veit und Comp, 1846, 259–499.

Frege, G., 1892, *Über Sinn und Bedeutung. Zeitschrift für Philosophie und philosophische Kritik*, NF 100. 1892, 25–50.

51 Wittgenstein 2000, Item 107, 257.

Gaffal, Margit, 2015: Forms of Life as a Philosophical Concept. In: *Emergent Forms of Life in Anglophone Literature. Conceptual Frameworks and Critical Analyses*, M. Basseler, D. Hartley, A. Nünning (eds.), Wissenschaftlicher Verlag, Trier, 27–37.

Goethe, J. W. Von, 1977: *Goethes Werke: Dramatische Dichtungen* III, Volumen 5. Torquato Tasso. München, C. H. Beck.

Husserl, E., 1968, *Logische Untersuchungen. Untersuchungen zur Phänomenologie und Theorie der Erkenntnis*, II/1. Tübingen, Max Niemeyer.

Lessing, G. E., 1996: *Werke in 8 Bänden*, H. G. Göpfert (ed.). Darmstadt, WB.

Padilla Gálvez, J. (ed.), 2016, *Action, Decision-Making and Forms of Life*. Berlin, Boston, Walter de Gruyter Verlag.

Padilla Gálvez, J. and M. Gaffal (eds.) 2011, *Forms of Life and Language Games*, Ontos Verlag, Frankfurt a. M.

Russell, B. 2003: The Basis of Realism, (1911), in: *Russell on Metaphysics*, ed. S. Mumford. London: Routledge, 87–90.

Searle, J. R., 1983, *Intentionality. An Essay in the Philosophy of Mind*. Cambridge, Cambridge University Press.

Ueding, G. (ed.), 1994, *Historisches Wörterbuch der Rhetorik*. Vol. 2, Tübingen, Niemeyer.

Voss, I. H., 1806, *Homers Odyssee. XIII -XXIV*. Tübingen, Gesang. Cotta'sche Buchhandlung.

Wieland, C. M., 1818, *Geschichte des Agathon*. Wien, Anton Doll.

Wittgenstein, L., 1969, *The Blue and Brown Books*. Oxford, Blackwell Publishing.

Wittgenstein, L., 1979, Notebooks 1914 – 1916, eds. by G. H. von Wright and G.E.M. Anscombe. Oxford, Basil Blackwell.

Wittgenstein, L., 2000, *Wittgenstein's Nachlass.The Bergen Electronic Edition. Normalized Transcription*. Oxford, Oxford University Press.

Wittgenstein, L., 2001, *Philosophische Untersuchungen. Kritisch-genetische Edition*. Ed. J. Schulte in collaboration with H. Nyman, E. von Savigny and G. H. von Wright. Frankfurt am Main, Suhrkamp.

Wittgenstein, L., 2009, *Philosophische Untersuchungen / Philosophical Investigations*. Transl. by G.E.M. Anscombe, P. M. S. Hacker and J. Schulte. Revised fourth edition by P. M. S. Hacker and J. Schulte. Oxford, Wiley Blackwell.

Severin Schroeder

Reasons and First-Person Authority

1 Introduction

Typically, an explanation of an event cites the causes that made it happen. According to what has been called the 'standard view' in the philosophy of action,[1] intentional human actions are just a subclass of events and the agent's reasons are the causes of their actions. In the 1930s this widespread causalism was disputed by Ludwig Wittgenstein, whose alternative view that reason-giving explanations should *not* be seen as a type of causal explanations, held sway in the 1960s. It was defended by Elizabeth Anscombe,[2] Peter Winch,[3] and A. I. Melden,[4] but was then forcefully challenged by Donald Davidson, in his influential paper 'Actions, Reasons, and Causes',[5] which by the 1970s had re-established causalism about reasons as the standard view. The standard causalist view, however, has generally been acknowledged to be troubled by a major problem: the problem of deviant causal chains.[6] Davidson himself had confessed not to be able to resolve this problem,[7] yet many of his followers remained more optimistic.

A more recent causalist view, propounded for example by Carlos Moya[8] and John Hyman,[9] is that the flaw both in Davidson's version of causalism and in the anti-causalism of Wittgenstein and his followers was their commitment to a Humean account of causation, as a law-governed relation between events. On a more flexible construal, Moya and Hyman argue, causalism can be defended against its critics. – In this paper, I shall, after a few preliminary clarifications, re-consider the deviant causal chain problem, with those more flexible causalist ideas in mind, and then try to give a defence of Wittgenstein's key anti-causalist idea: first-person authority about reasons.

1 Cf. Sandis 2009, 2.
2 Anscombe 1957.
3 Winch 1958.
4 Melden 1961.
5 Davidson 1963.
6 Moya 1990, 114–28.
7 Davidson 1973, 79.
8 Moya 2014, 200–4.
9 Hyman 2015, ch. 5.

https://doi.org/10.1515/9783110560282-129

2 Reasons and causes

To begin with, let us note that reasons are not causes: the two concepts are categorially different. Although a thought may well be cited both as a reason and as a cause, this is due to a well-known ambiguity: by a thought we can mean either the occurrence of a thought (the mental event of thinking of something at a given moment) or the contents of the thought (what we think). Thus, the mental event of my thinking or remembering that I've run out of cash may be said to *cause* me to set off in search of a cash machine, whereas the *reason* I offer to justify my behaviour is the content of my thought: that I need some cash.[10] It can occasionally happen that my reason for acting is not *what* I think, but the fact that a certain thought occurs to me. This is nicely illustrated by a pair of examples given by Hyman:[11]

> (1) Roger fled to Brazil because he believed that he was being pursued by the Security Services.
> (2) Roger saw his doctor because he believed that he was being pursued by the Security Services.

In the second case, on a natural interpretation, Roger's action was not motivated by the contents of his belief, but by its occurrence, which he took to be pathological. However, even in such a case reason and cause are not identical. The occurrence of such irrational thoughts caused Roger to seek medical advice, but the reason for his action was not the (datable) *occurrence* of the thought, but the (non-datable) *fact* that it occurred.[12] The fact that a football match lasts 90 minutes does not itself last 90 minutes, and if a stone hit the window it is not *the fact that a stone hit the window* that hit the window (and caused it to break). Similarly, *my thirst* may have caused me to drink a whole pint, but the reason I did so was *that I was thirsty*: a fact, not a desire.

This categorial difference between reasons and causes does of course not rule out a causalist construal of reason-giving explanations. But instead of saying that reasons are causes, a causalist should propose, more carefully, that a reason-giving explanation is correct if and only if a corresponding causal claim is correct. For example, on the causalist view, the following two statements are equivalent (subject perhaps to some minor qualifications):

---

10 Wittgenstein *PLP*, 123; Gnassounou 2014, 58.

11 Hyman 2001, 182.

12 Gnassounou 2014, 59.

(3) I went in search of a cash machine because I had run out of cash.
(4) The thought that I had run out of cash caused me to go in search of a cash machine.

3 The problem of deviant causal chains

However, the causalist view faces the problem of deviant causal chains. Two versions of this problem need to be distinguished.

(i) First, in the well-known examples given by Roderick Chisholm[13] and Davidson,[14] the resulting action although caused by the right belief and desire is not even an intentional action. Rather, the agent's thoughts make him nervous and thus cause him to bring about the envisaged action unintentionally (e.g., careless driving makes him inadvertently and coincidentally run over the very person he is thinking of murdering). This kind of case serves to highlight that the existence of a suitable desire-belief pair is not sufficient for the intentional action in question to be carried out. I may want to do it and know how to do it – but not be able to bring myself to do it. Therefore, causalists proceeded to insist that the mental cause of an intentional action must be an *intention* to carry out that action.[15]

However, this modification does not overcome the deviant causal chain problem. Assuming for the moment that an intention is indeed a mental occurrence suitable for causing an action (perhaps envisaging it as a kind of decision to act),[16] it may well lead to the action being carried out unintentionally. In Chisholm's example of the homicidal driver, we can imagine that the man was already determined and on the way to committing the crime when his eagerness to carry it out made him kill his victim accidentally. It is evidently not enough for the intention to figure among the causes if in the end it doesn't inform the action itself. As Moya puts it, in his critical discussion of Irving Thalberg's causalist theory: it is not enough that as a result of an intention an agent fulfils that intention; 'the agent has to be conscious of fulfilling his intention'[17]. That is indeed the crucial point: for an action to be intentional the agent has to be aware of carrying it out. But this awareness is a feature of the action itself and

13 Chisholm 1966, 29–30.
14 Davidson 1973, 79.
15 Brand 1984, 31–5; Thalberg 1984, 256; Stout 2010.
16 For reasons to think otherwise, see Anscombe 1983; Rundle 1997, 197–206; and Clot-Goudard 2015, 68–76.
17 Moya 1990, 122.

not a causal factor preceding it. It is certainly possible that a plan or intention precedes the action, but that is neither a sufficient nor a necessary condition of intentionality. It is quite enough for the intention to be there in the act, as an ability I have of explaining what I'm doing. The causal preliminaries cancel out as conceptually irrelevant.

(ii) Another version of the deviant causal chain problem occurs when the intentionality of the action is not in question, but we want to ascertain the reason for which the person acted. This is the point where Davidson presents his only positive argument for a causalist construal of reasons. Where a person performs a certain action and is aware of more than one reason to do so, we need a criterion by which to decide which of the reasons did in fact motivate the person to act. For Davidson, the only plausible criterion to single out the (actual) *reason why somebody ⍰-ed* from (potential) *reasons to ⍰* is a causal link between reason and action.[18] But, as Davidson had to acknowledge, it is easy to think of cases in which an action is caused by the thought of a reason that was not the one for which the person acted. In a fairly common scenario, considering reason *A* may lead to the discovery of reason *B* which then motivates me to act. The thought of reason *A* would have caused the action, although it wasn't the reason for which I acted. An obvious response would be to demand that the required causal link between the thought of the reason and the action must not be indirect, but proximate.[19] But that doesn't help, for we can easily imagine that the proximate consideration that triggers the action provides a mere reminder or pretext, not the reason for which the person acts. For example, hearing a moving account of the economic difficulties of the local bicycle shop leads me to buy a bicycle there; yet I may have been planning to make this purchase anyway for selfish reasons and now I have just been reminded of this intention. Perhaps I do indeed like the idea of supporting a local shop, but it would never have been a reason for me to buy a bike had I not wanted to do so anyway.

Others have suggested that the (thought of the) reason must be required to cause the action 'in a way that is characteristic of' acting for that reason,[20] or in the way in which a disposition causes its manifestations.[21] But it is hard to see how that could provide us with a criterion for deciding, in a given case, whether somebody acted for reason *A* or for reason *B*. One suspects that what this kind of uninformative restriction really amounts to is that we do not in fact rely on a

---

18 Davidson 1963, 9.
19 Thalberg 1984, 257.
20 Stout 2010, 162.
21 Hyman 2015, 116–127.

causal analysis for identifying agential reasons. Rather, when we think we know independently that somebody acted for reason *A* and not for reason *B*, we say post hoc that only (the thought of) reason *A* caused the action 'in the right way'. In other words, causalism, unhelpfully qualified, becomes a mere postulate that doesn't do any work.

The truth is that even if somebody did specify what the 'right' causal pathway was supposed to be, we know far too little about the causal pathways of our thoughts and desires for such a causally construed concept of an agential reason to be usefully applicable. The insufficiency of our knowledge of the causal impact of thoughts and desires tends to be disguised by the fact that we focus on a few examples where the causal pathways, and the strength or weakness of particular desires are simply assumed to be obvious. Yet psychological causes are not like inner billiard balls whose paths can be followed by introspection. This may to some extent be the case with mental occurrences: I can, at least sometimes, fairly reliably report a causal succession of ideas (– how a given sentence reminds me of my sister – which memory in turn brings up a mental image of my school – which then reminds me of the book on architecture I just bought – which then makes me pick up that book). But dispositional phenomena, such as beliefs and inclinations, are not mental occurrences and so their effects cannot be read off their temporal position.

Acknowledging our ignorance in this area, Alvin Goldman suggested that an 'explanation of how wants and beliefs lead to intentional acts would require extensive neurophysiological information'[22]. But for one thing, it is far from clear that in a case where the agent is aware of various potential reasons neurophysiological data could neatly decide which considerations were the principal cause of his action. For all we know, even thoughts that we don't regard as relevant may play an important role in the complicated causal network of our psyche and its underlying physiology. For another thing, our concern is not with a possible future concept of a reason, defined in the light of future scientific findings and possibilities, but with a concept that has played a central role in our discourse for centuries: for as long as people have been articulate enough to inquire about and explain their behaviour to each other. Clearly, relying on criteria that remain either unspecified or that we are in fact unable to apply, causalism fails to offer a plausible analysis of this everyday concept.

---

22 Goldman 1970, 62.

4 Reasons and first-person authority

At the core of Wittgenstein's distinction between reasons and causes lies the idea that the grammar of the former is characterised by first-person authority:

> ...a person does not generally know the *causes* of his activities. ...he will frequently be mistaken in specifying the cause. Strangely enough, he cannot be mistaken in specifying his *reason*. ... That is, we *call* the reason that which he gives as his reason. The cause of an action is established by observation, namely *hypothetically*, i.e., in such a way that further experiences can confirm it or contradict it.[23]

John Hyman sees in this (and a similar passage in *BB* 15) only the claim that causal explanations are *conjectures*, whereas we are immediately certain of our reasons,— to which he rightly objects that causal statements can be certain and immediate too.[24] In fact, that is something Wittgenstein himself remarked and illustrated by the example of being startled by suddenly seeing a light: in which case one does not *hypothesise* on the basis of repeated observations about the cause of one's bodily movement, but is immediately aware of that cause.[25] So it is indeed inaccurate to say that 'one can only *conjecture*' the cause (as Wittgenstein does in *BB* 15), but that is not his main point.

The crucial contrast is not between immediate knowledge and mere conjecture based on repeated observation, but between, on the one hand, an authority conferred by our concepts (hence not really a matter of knowledge at all) and, on the other hand, any kind of empirical finding, be it immediate perception or a tentative conjecture. In the latter case there is an independent state of affairs of which I come to know, be it step by step through repeated observation or immediately. Either way, my cognition is fallible. Error may by highly unlikely or even practically impossible, but extraordinary circumstances that would prove my judgement erroneous are at least conceivable. Thus, it is at least imaginable that the car I see right in front of me in broad daylight proves to be an ingeniously constructed hologram. Similarly, when I suddenly see a light or a face and startle, it is conceivable (however far-fetched and unlikely) that my startling reflex was triggered by some electrodes in my brain quite independently of what I happened to see at that moment.

---

23 Wittgenstein *VW*, 109–11.
24 Hyman 2015, 118.
25 Wittgenstein *CE*, 408.

By contrast, first-person authority does not underwrite reports about independent occurrences, but marks certain kinds of sincere avowals of a person's impression as constitutive of a certain concept. Thus, when I can sincerely complain that something feels painful to me, then – by definition or by grammar – it *is* painful. This concept of feeling does not allow any room to a distinction between appearance and reality; that is, error in this case is logically impossible. (Not because I know so reliably whether I'm in pain, but simply because this is not really a case of knowledge at all.[26])

Beliefs and preferences provide other examples of first-person authority. Whereas my *considered* opinions and *long-term* preferences may require some working out (researching and weighing of evidence, testing how different things agree with me), there are also concepts of current belief and spontaneous preference that capture how things appear to me and what I am inclined to choose at a given moment. There, again, error is logically excluded. If I can sincerely say that I don't think Arsenal will win the Champions League and that I'd like to have a cup of tea now, then that is *ipso facto* what I believe and what I want at that moment.

Similarly, according to Wittgenstein, there is a concept of a person's reason that is simply constituted, by that person's possible sincere avowal of his reason. Not because we take people to be so reliable at finding out what their reasons are, but because a reason (in this sense), unlike a cause, is not an independent occurrence at all: it is simply defined to be what an agent can honestly say in response to the question: 'Why are you doing it?'. Thus, Wittgenstein's account offers a straightforward solution to the problem that stymied Davidson: the problem of identifying the reason for which somebody acted.

To see more clearly how first-person authority is built into our concept of an agential reason, consider the concept of an intentional action. Intentional actions are a subclass of conscious voluntary actions. They are 'actions that are chosen either as ends in themselves or as means to some other ends'.[27] The important thing to note is that to be conscious of an intentional action – to know what it is one is doing – is to be conscious or aware of its intentional characteristics, which make it the action it is. In the case of an intentional action, to be aware what one is doing must include being aware that one is doing it as an end in itself or as a means to some other end. In other words, it conceptually includes an awareness of one's reasons, i.e. first-person authority.

---

26 Wittgenstein *PI*, § 246; Schroeder 2006, 208–10.
27 Kenny 1989, 45.

Why, for example, am I pushing a plastic card into a slot and pressing a few buttons? My reason for doing so (— I'm making a payment, buying a bicycle, with which I'd like to go on weekend bike rides —) is an essential characteristic of my very action. Instead of 'Why are you pressing those buttons?' one might just as well ask: 'What are you doing there pressing those buttons?'. My reason is not just describing the preliminaries that led to the action (as causalism suggests), it characterises the action itself. If I weren't able to explain my action's intentional features (by either giving a reason or, less likely, saying that I thought pressing buttons was worthwhile in itself), then it just wouldn't qualify as an intentional action. It might just be a kind of fidgeting.

Note that first-person authority in giving one's reasons — as all kinds of first-person authority — is strictly speaking only a matter of present tense. That as an intentional agent I can authoritatively say what I am engaged in doing does obviously not guarantee the truth of my (sincere) descriptions of my past actions. However, with regard to one's actions in the recent past where, given ordinary human memory capacities, one can be trusted to remember the relevant circumstances of the situation, first-person authority is also granted retrospectively. That is because, as long as a person's views and preferences haven't changed, we can assume that the justification he offers now is probably not much different from the one he would have offered at the time of the action.

Giving an explanation of one's recent action is to perform another action, often manifesting the same want and belief as the original action. To explain one's verbal utterance is normally a matter of paraphrasing or supplementing it. 'It's a disgrace!' —?— 'I mean: it's very unfair that he got sacked after all he has done for the company.' In this case the philosophical question as to the virtual infallibility of an agent's explanation of his action is easily answered. Action and explanation are on the same footing. In both cases the person speaks his mind; and there is no puzzle about our ability to say repeatedly what we think about a subject, being a little more explicit the second time.

There are similar pairs of actions in the non-verbal sphere. Suppose you see Fanny Robin throwing a snowball against a wall, and ask yourself: 'What on earth did she do that for?' Then you see her throwing another snowball against a first-floor window, and you understand that that is what she was trying to do the first time: hit the window to attract someone's attention. The second action explains the first by pursuing the same purpose more expediently, and thus more evidently.

Now consider the case where being asked for her reason after the first throw she explains: 'I want to hit that window to attract Sergeant Troy's attention'. Her ability correctly to give her reasons ceases to look puzzling if we see this kind of case as a variation upon the preceding ones. Action and explanation are simply two actions manifesting the same intention (e.g. to attract someone's attention) and the same belief (e.g. that it can be done by throwing a snowball against a window). Given that she can speak, her ability to say for what reason she acted is not more surprising than her ability to repeat the action.

Here we have an illustration of how giving one's reason is not a matter of describing what happened in one's mind. One only has to express truthfully what one wants and believes (or knows) in the matter. Of course, things are a little different if one's wants or beliefs are no longer the same. Once Fanny Robin has succeeded in hitting the window and thus attracted Sergeant Troy's attention, she will no longer have a want to do so. Hence, explaining her own behaviour will no longer be merely a matter of expressing a *current* want and belief. But even then no more is required of her than a sincere expression of her inclinations, only in a conditional form: 'If I hadn't yet succeeded in hitting the window, what would I want to do?' Again, the answer that will explain her action need not draw on any mental events she might remember; she only has to express truthfully her present inclinations to say what under such circumstances she *would* want to achieve and believe to be achievable by the behaviour in question.

However, when it comes to explaining our behaviour months or years ago we are obviously fallible. I may not remember what my wants, interests and plans were a few months ago, so that the reasons I now attribute to myself retrospectively may not be the ones I had at the time (as others remembering my earlier explanations may know). And we are also fallible with respect to the history and development of our long-term intentions and reasons. Past intentions that led to an action over a longer period of time need not be part of the intentional characteristics of the action itself. – Suppose pity with the plight of our local bicycle shop owner gives me the idea of supporting him by buying a bike, even though I never had any thoughts of using a bicycle. But once I decided on buying a bike from him, I naturally started thinking what I could do with it. The idea of weekend bike rides arose and pleased me so much that by the time of the purchase I regard it as the only reason for my action. It may be that I've forgotten the original motive of compassion, but it may also be that I regard it as a change of mind: two months ago I might have bought a bike in order to support the shop, but now I don't feel particularly concerned about the shopkeeper anymore and give only my weekend plans as my reason. The original reason figures in the causal history of my action, but is not the reason for which I acted.

5 The 'real' reason

There are various situations in which we are not satisfied with the reasons given by an agent for their action, situations in which we are likely to dig deeper for (what we may be inclined to call) the 'real' reason. Does that contradict Wittgenstein's account of first-person authority about reasons? – Five kinds of cases can be distinguished:

5.1 People can be dishonest about their reasons

It often happens that one suspects a person's stated reason not to be their real reason, simply because they may not have been entirely honest. Such suspicions are typically based on a perceived inconsistency between the reason stated and the views and preferences the person manifested by their behaviour on other occasions. For example, a ruthlessly selfish businessman's donations to charity may be suspected to be given merely for prudential reasons, to improve his image, even if he solemnly declares that he only wanted to alleviate suffering. That is straightforward and in no way in conflict with the view that reasons are what agents can *truthfully* present as their reasons.

But is it not possible that even a *sincere* statement of one's reasons is contradicted by one's behaviour? Here two kinds of cases need to be distinguished. First, an intentional action and the reasons it involves may, of course, be out of character. Human characters are rarely entirely homogenous. Even a money-grabbing egoist may have moments of altruistic generosity. Secondly, we could try to imagine a more serious inconsistency. Suppose somebody is honestly convinced to act for a reason, a belief, which his surrounding behaviour shows him not to have. For example, Jones turns down the music at 11 pm in order not to disturb the neighbours – although his knowledge that the neighbours are away is shown by the fact that he regularly enters their flat to feed the cat and water the flowers during their absence. Even that would not be particularly uncommon, as the knowledge may have momentarily slipped his mind, so that when turning down the music he was indeed sincerely concerned not to disturb the neighbours. If, however, we assume that the two contradictory beliefs are being manifested more or less simultaneously, the ascription of either becomes problematic. The person's behaviour, verbal or non-verbal, would just be too confused for our concepts of belief and reason to get a firm foothold. A certain consistency in our verbal and non-verbal expressions of beliefs has to be presupposed for it to be possible to attribute reasons to people, just as it is a precondition of our attributing beliefs to them.

5.2 People can be unclear about their reasons

I may feel unable to give a clear answer to the question why I bought a bike. I may be aware of a number of reasons – considerations speaking in favour of it that I find plausible – without being prepared to single out one of them as decisive. Then it just has to be accepted that there is no single reason why I did it.

5.3 Agential reasons can be (patently) erroneous

As Jonathan Dancy puts it: reasons are non–factive.[28] The reason for which I acted may be that *p* (e.g., that the bicycle I bought was of good quality) – even if it is not in fact the case that *p* (– I was wrong in thinking that the bicycle was of good quality). In such a case, where the agent is mistaken, or even deluded, our interest may well shift from a reason-giving explanation to a causal explanation. Yet it would be a conflation of categories to say that, in such a case, the unknown cause was the agent's 'real reason'.

For an example of such a conflation, consider the following psychological experiment:[29]

> ...passers-by were invited to evaluate items of clothing – four different nightgowns in one study (378 subjects) and four identical pairs of nylon stockings in the other (52 subjects). Subjects were asked to say which article of clothing was the best quality and, when they announced a choice, were asked why they had chosen the article they had. There was a pronounced left-to-right effect, such that the rightmost object in the array was heavily chosen. For the stockings, the effect was quite large, with the right-most stockings being preferred over the leftmost by a factor of almost four to one. When asked about the reasons for their choices, no subject ever mentioned spontaneously the position of the article in the array.[30]

The psychologists Richard Nisbett and Timothy Wilson interpret this to show that people tend to be ignorant of the real reasons of their actions.[31] But, as Con-

---

28 Dancy 2000, 134.
29 I owe this example to Constantine Sandis 2015.
30 Nisbett and Wilson 1977, 233.
31 Nisbett and Wilson 1977, 249.

stantine Sandis has pointed out,[32] Nisbett and Wilson just fail to distinguish two different concepts: that of an agent's reason – what the agent is prepared to offer as a justification of his action – and that of the causal factors influencing an action.

The conflation is partly due to the fact that the word 'reason' can also be used to denote a cause (rather than an agential reason). Thus it is perfectly acceptable to say:

(5) The reason that so many subjects chose those nylon stockings is that they were placed in the rightmost position.

Just as one can plausibly say:

(6) The reason that he proposed to her that night was that he was drunk.

Here, instead of 'the reason' one could also say: 'what explains the fact'. By contrast, an agential reason would be indicated by prefacing the word with a possessive pronoun (*their* reason, *his* reason). Yet it would take a very odd character to make the following statement true:

(6a) His reason for proposing to her that night was that he was drunk.

It is much more probable that the man's reason for proposing to the lady in question was that he thought her a desirable wife, while the drunkenness just allowed him to overcome his usual shyness; or alternatively, it may have clouded his judgement. In the first case, the drunkenness caused him to act on his reason; in the second case, the drunkenness induced him to have a new (and probably ill-judged) reason (namely, to find the lady suitable as a wife).

Similarly, consumer psychology cunningly identifies causal factors, such as location or packaging, that make people believe certain products to be of superior quality. The reason for their choice, then, is what they mistakenly belief – not the marketing trick that caused them to have such a belief. Again, the following would describe a rather odd scenario:

(5a) Miss Jones's reason for choosing those nylon stockings was that they were placed on the extreme right.

---

32 Sandis 2015, § 3.

We would be entitled to ask her why she thought that it was preferable to take stockings from the right side of the shelf. (Is she superstitious? Or has she heard of a convention that that is where shopkeepers tend to put the best products?)

In short, what psychologists, such as Nisbett and Wilson, show is not that people are mistaken about their reasons, but simply that they are often ignorant of the causes of their having, and acting on, certain reasons. That what they believe and take as their reason is false – that they have been duped into believing it by factors unknown to them – doesn't mean that it isn't their real reason. It only means that in such a case the reason-giving explanation itself stands in need of a (further) explanation, which is likely to be a causal explanation. For given the falsity of the reason (that in fact the preferred stockings were *not* of better quality), we want to know why the agent held such a false belief.

5.4 The action to be explained was not in fact voluntary.

The most extreme cases of delusional reasons ('confabulation') are those that involve post-hypnotic action. Wittgenstein considers the following case:[33]

> The experimental subject, who has under hypnosis been given a particular task to execute, e.g. to put up his umbrella, does precisely this – but has no inkling why he does what he does; asked to account for it, he may well invent a motive – 'I wanted to check that the umbrella is in good shape', and believe in it perfectly sincerely; all the same he is mistaken.[34]

However, what according to Wittgenstein the experimental subject is mistaken about is the *cause* of his action, not his reason.[35] The action was not caused by a preceding thought that the umbrella might be defective, yet we can still accept his stated reason as the best justification of his behaviour he can think of. Of course, under the circumstances, we are not much interested in such a justification, since we know that his behaviour wasn't voluntary. What he gave us may still be regarded as *his reason* – but what he did was not *his intentional action*, nothing that could be explained by the agent's reason. What happened here was, in fact, the hypnotist's doing, and so we might be more interested in the hypnotist's reason for it.

---

33 It should be noted that in this passage Wittgenstein uses the word '*Motiv*,' translated as 'motive', for 'what somebody specifies as a reason for his action' Wittgenstein *VW*, 424–5.
34 Wittgenstein *VW*, 424–5.
35 Wittgenstein *VW*, 424 f.

(v) Agential reasons may be found insufficient to justify the action (even in light of the agent's views and preferences).

There are certain cases of self-deception where the agential reason given is not false or deluded, but does not appear sufficiently strong to justify the action.

Of course somebody's reasons can appear feeble to us simply because we don't share their views and preferences. I should never regard the concert of a Heavy Metal Band in London as a good reason for going there, but I should not for that matter question others' explanation that that was their reason for going to London taking into account that they may have weird tastes in music. But there are other cases where a stated reason appears insufficiently weighty even by the agent's own ideas and standards.

I may, for example, sincerely give as a reason for my telling someone off his lack of politeness. Yet a careful observer who knows me sufficiently well realizes that the occasion was fairly trivial and that I would never have lost my temper so much, had it not been for the fact that I felt jealous of the person I told off, perhaps without being clear about it myself. Here, again, a reason sincerely offered by the agent is not to be contradicted. The *reason* for my action was indeed the one I gave (always assuming my sincerity, which might require that I reflect carefully on my relevant views and beliefs). However, since my behaviour was less under my control than I believed, my behaviour could not be fully accounted for by my reason. Part of it (the inappropriate vehemence of my outburst) could only be explained causally: by my being carried away by an emotion I hadn't yet fully taken stock of.

The domain of reason-giving explanations is reasonable behaviour. Hence, to the extent to which an action is unreasonable we will look for a *causal* explanation, to supplement, or in some cases even to replace the reason-giving explanation.

6 Conclusion

As argued, following Wittgenstein's ideas, it is a mistake to think that reason-giving explanations are just agents' attempts at causal explanations. The concept of an agential reason is crucially different from that of a cause (— just as a person's avowal of how much they value something must not be misconstrued as their attempted estimate of the thing's objective monetary value). Nevertheless, many reasons can also be said to identify causes. I stopped because the traffic lights turned red. That is my reason, but it is equally correct to say that the red lights caused me to stop. Again, if I develop a desire to own an *Opel*

Insignia this desire may lead (cause) me to buy such a car; yet I may also give as the reason for my purchase that I desired to own such a car.

The upshot of Wittgenstein's account is not that reasons cannot refer to causal factors, but merely that the *concept* of a reason is not the *concept* of a cause. Due to this conceptual difference, there is considerable overlap, but no complete correspondence between reason-giving and causal explanations of an action. That is, for one thing, because not all reasons refer to independently existing psychological occurrences or states.[36] The second fact responsible for divergences between the two concepts is the one discussed in this paper, namely that reasons, unlike causes, are covered by first-person authority. It means that although I may be infallible in giving certain facts or circumstances as a reason for my action, I do not enjoy the same infallibility with respect to the causes of my behaviour. Although normally when I stop in front of a red traffic light, the red light provides both the reason and the cause for my stopping, it is conceivable that my stopping had different causes: perhaps I hallucinated the traffic lights or it was something else (say, hypnosis) that caused me to brake. In such a case I would still be correct to give the red lights as my reason for stopping.

But first-person authority is relevant not only in cases of misperception or error. It also allows the agent to single out circumstances as decisive from a justificatory point of view when they play only a minor role, or perhaps no role at all, from a causal perspective; while, on the other hand, an observed circumstance that was causally crucial, getting the action under way, may rightly *not* be regarded by the agent as his reason for acting.[37]

7 References

ANSCOMBE, G.E.M., 1957: *Intention*. Cambridge/Mass., Harvard Unnversity Press, 1963.

ANSCOMBE, G.E.M., 1983: 'The Causation of Action', in: M. Geach & L. Gormally (eds), *Human Life, Action, and Ethics*. Exeter, Imprint Academic.

BRAND, M., 1984, *Intending and Acting*. Cambridge/Mass., MIT Press.

CHISHOLM, R.M., 1966, 'Freedom and Action', in: K. Lehrer (ed.), *Freedom and Determinism*. New York, Random House, 11–44.

CLOT-GOUDARD, R., 2015, *L'explication ordinaire des actions humaines*. Montreuil-sous-Bois, Ithaque.

DANCY, J., 2000, *Practical Reality*. Oxford: Oxford University Press.

36 Rundle 1997, 12, 168; Schroeder 2001, 159–63.

37 I am grateful to Casey Doyle for his comments on this paper.

DAVIDSON, D., 1963, 'Actions, Reasons, and Causes', in: *Essays on Actions and Events*, Oxford, Clarendon Press, 1980, 3–19.
DAVIDSON, D., 1973, 'Freedom to Act', in: *Essays on Actions and Events*. Oxford, Clarendon Press, 1980, 63–82.
GNASSOUNOU, B., 2014, 'Intention et action', in: R. Clot-Goudard (ed.), *L'explication de l'action: Analyses contemporaines*. Paris, Vrin, 57–78.
GOLDMAN, A.I., 1970, *A Theory of Human Action*. Princeton, Princeton University Press.
HYMAN, J., 2001, 'Knowledge and Self-knowledge', in: S. Schroeder (ed.), *Wittgenstein and Contemporary Philosophy of Mind*. Basingstoke, Palgrave; 171–193.
HYMAN, J., 2015, *Action, Knowledge, & Will*. Oxford, Oxford University Press.
KENNY, A., 1989, *The Metaphysics of Mind*. Oxford, Clarendon Press.
MELDEN, A.I., 1961, *Free Action*. London, Routledge & Kegan Paul.
MOYA, C. J., 1990, *The Philosophy of Action*. Cambridge, Polity Press.
MOYA, C. J., 2014, 'Les aspects causaux de l'explication de l'action', in: R. Clot-Goudard (ed.), *L'explication de l'action : Analyses contemporaines*. Paris, Vrin, 187–206.
NISBETT, R.E. and T.D. WILSON, 1977, 'Telling More Than We Can Know: Verbal Reports on Mental Processes', *Psychological Review*, Vol. 84, No. 3, 231–259.
RUNDLE, B., 1997, *Mind in Action*. Oxford, Clarendon Press.
SANDIS, C. (ed.), 2009, *New Essays on the Explication of Action*. London, Palgrave Macmillan.
SANDIS, C., 2015, 'Verbal Reports and "Real" Reasons: Confabulation and Conflation', in: *Ethical Theory and Moral Practice*, 18:2, 267–280.
SCHROEDER, S., 2001, 'Are Reasons Causes? A Wittgensteinian Response to Davidson', in: S. Schroeder (ed.), *Wittgenstein and Contemporary Philosophy of Mind*. Basingstoke: Palgrave, 150–170.
SCHROEDER, S., 2006, Wittgenstein: The Way Out of the Fly-Bottle. Cambridge, Polity.
STOUT, R., 2010, 'Deviant Causal Chains', in: T. O'Connor & C. Sandis (eds), *A Companion to the Philosophy of Action*. Chichester, Wiley-Blackwell, 159–165.
THALBERG, I., 1984, 'Do our intentions cause our intentional actions?', *American Phil. Quarterly* 21, 249–260.
WINCH, P., 1958, *The Idea of a Social Science*. London, Routledge & Kegan Paul.
WITTGENSTEIN, L., 1958, *The Blue and Brown Books* (BB). Oxford, Blackwell.
WITTGENSTEIN, L., 1965, *The Principles of Linguistic Philosophy* (PLP), by F. Waismann, based on Wittgenstein's dictations, edited by R. Harré. London, Macmillan.
WITTGENSTEIN, L., 1993, 'Cause and Effect: Intuitive Awareness' (CE), in: *Philosophical Occasions 1912-1951*, eds: J. Klagge & A. Nordmann. Indianapolis, Hackett, 370–426.
WITTGENSTEIN, L., 2003, *The Voices of Wittgenstein. The Vienna Circle* (VW), by Ludwig Wittgenstein and Friedrich Waismann, edited by Gordon Baker. London, Routledge.
WITTGENSTEIN, L., 2009, *Philosophical Investigations* (PI), Fourth Edition. Edited by P. M. S. Hacker and J. Schulte. Translated by G. E. M. Anscombe, P. M. S. Hacker and J. Schulte. Oxford, Wiley-Blackwell.

Amir HOROWITZ

On the Very Idea of a (Natural) Intentional Relation

I

In an often quoted passage, W.V.O Quine writes:

> One may accept the Brentano thesis [of the irreducibility of intentional idioms] either as showing the indispensability of intentional idioms and the importance of an autonomous science of intention, or as showing the baselessness of intentional idioms and the emptiness of a science of intention. My attitude, unlike Brentano's, is the second.[1]

According to Fodor, "If aboutness is real, it must be really something else."[2] This sentence (also often quoted) – call it "Fodor's dictum" – means that realism about intentionality entails that the intentional can be reduced to the "natural". Fodor indeed believes that intentionality is "something else". Quine, as we saw, rejects this idea – he endorses "Brentano thesis". But unlike Brentano, Quine accepts Fodor's dictum. His anti-reductionism coupled with his accepting Fodor's dictum led him to the view of "the baselessness of intentional idioms", that is, to intentional irrealism.

For everyone who accepts Fodor's dictum, the viability of intentional naturalistic reductionism is all important. My aim in this paper is to show that there is something basically flawed with the notion of "natural" intentional relation and thus with the very idea of a naturalistic reduction of intentionality. Since I accept Fodor's dictum, I take the baselessness of this idea to mean that intentional realism is false. Of all three philosophers, then, I believe that Quine has the upper hand. But the present paper is devoted to undermining the idea of a naturalistic reduction of intentionality, and its case against intentional realism is conditional upon the truth of Fodor's dictum.

Intentional realism is the thesis that various concrete entities (prominently – various mental states) have intentional properties. According to intentional

1 Quine 1960, 221.
2 Fodor 1987, 97.

https://doi.org/10.1515/9783110560282-145

irrealism, virtually no concrete entity ever has an intentional property.[3] It may be useful to put this idea in terms of content ascriptions. In these terms, intentional realism is the thesis that most – or a significant percentage – of our content ascriptions are true, while on intentional irrealism (in a strong sense) all or at least most content ascriptions are false – there is a systematic error in the practice of content ascription. According to an even stronger version of intentional irrealism, the one I endorse, intentional ascriptions can be neither true nor false: they all lack truth conditions. We may call this version "the semantic conception of intentional irrealism", and its denial – "the semantic conception of intentional realism".[4]

The talk of truth conditions of content ascriptions enables us to highlight the conditional nature of the purpose of this talk from a different direction: it aims to show that content ascriptions made from the third-person perspective lack truth conditions, and thus to suggest a case for intentional irrealism that is conditional on the truth of the claim that content ascriptions made exclusively from the first-person perspective lack truth conditions. Indeed, I believe that this is so – I take such ascriptions to be vacuous, but I do not argue for this idea in the present paper.[5]

Intentionality seems to many philosophers to be an elusive phenomenon, perhaps a mysterious one. It isn't clear how one entity can "transcend" itself and be about, or represent, another entity.[6] The essence of the naturalistic-reductionist approach to intentionality is that one entity represents another simply by (or in the sense of) being related to the other entity by means of a natural relation.[7] The "transcendence" of intentionality on this approach comes down to be a simple relation of a familiar kind: intentionality transcends its bearer no more than, say, the causal property of the kicking foot that connects it with the ball kicked by it. On this approach, the intentional relation – the rela-

3 Defined in relation to *concrete* entities, intentional irrealism is not committed to denying propositions, which are intentional by their very definition.

4 I prefer the terminology of irrealism over that of eliminativism, first, since the latter has a methodological connotation which is irrelevant to the issue of the present paper, and second, since eliminativism is associated with the repudiation of mental states whereas my repudiation of intentionality does not commit me to such a view. I will explain this point in section IV.

5 I argue for this claim in Horowitz (in preparation).

6 Kant considered the question of how anything in the mind can represent anything outside the mind the most difficult philosophical question (see Kant's letter to Herz, Feb. 21, 1772, in Kant 1967). The characterization of intentionality in terms of transcendence is due to Husserl 1913.

7 The first explicit formulation of this approach known to me is Hartry Field's 1978.

tion of, e.g., thinking of X or believing that p – is not *sui generis*. It is a relation of a kind well-known to us, paradigmatically some kind of causal relation, or a causal relation that obtains under some specified conditions, etc. Richard Rorty said that the mind is related to the world causally but not semantically. The naturalist-reductionist may be taken to reply: the mind is related to the world causally *and therefore* semantically. Some natural, extensional, relation is all that is needed for ensuring semantics. So the naturalistic-reductionist approach may appear to dissolve the mystery of intentionality, which is the mystery of its transcending nature.

Before turning to show that the naturalistic-reductionist reply to the question of intentionality is inadequate, let me make a few clarificatory remarks about it. First, naturalistic approaches standardly have to face the challenge of characterizing "natural". But proponents of the idea of the naturalistic reduction of intentionality do not face this challenge, since this reduction simply aims to reduce the intentional to the *non-intentional* or *non-semantic*.[8]Essential to the reductionist-naturalistic approach to intentionality are the convictions that the world can be described non-semantically, and that its complete non-semantic description entails its (complete) semantic description. (Non-reductionist naturalists, such as me, reject the latter). Of course, many proponents of intentional reductionism are motivated by the desire to make room for intentionality in a purely physical world, and successful reduction of the intentional to the non-intentional would indeed mean that intentionality poses no special challenge to physicalism, but those convictions have no necessary connection to the issue of physicalism.

Second, the reduction that is the focus of this paper aims to reduce intentionality to a natural (non-semantic) *external relation*. Not all reductions of intentionality are relational, and not all relational reductions of intentionality

---

8 Some philosophers object to the naturalistic-reductionist program on the grounds that, first, intentionality is irreducibly normative (in this or that sense), and second, the notion of the natural is inapplicable to what is irreducibly normative (for various normativist approaches to meaning or content see, e.g., Boghosian 2008; Brandom 1994; Gibbard 2012; Kripke 1982). Adopting both claims makes the characterization of the naturalistic reduction of intentionality in terms of a reduction to the non-intentional insufficient. However, I reject the view that intentionality is irreducibly normative in any significant sense. I think Jerry Fodor's following objection to this view is decisive: To apply a term to a thing in its extension *is* to apply it correctly; once you've said what it is that makes the tables the extension of 'table's, there is surely no *further* question about why it's *correct* to apply a 'table' to a table. (Fodor 1990, 135-136n) So I think that proponents of the naturalistic reduction of intentionality need not be bothered by its alleged normativity.

employ external relations, that is, relations to the non-mental environment. Thus, accounts of intentionality in terms of consciousness, such as Searle's (1992), are non-relational reductions (though Searle might dislike the characterization of his account as reductionist),[9] whereas solipsistic conceptual role semantics such as Block's (1986) is relational but not external or environmental reduction.[10] (Theories that account for dimensions of content such as its functional role or inferential role but leave out the "hooking up to the world" dimension are not theories of intentionality, in my use of the term "intentionality".) Some examples of reductionist theories in the sense under discussion are informational theories, optimal conditions theories, asymmetric dependence theories, long-arm conceptual role theories, and teleological theories.[11]

Third, the naturalist-reductionist approach need not restrict itself to accounting for the intentionality of the mental and take the intentionality of language, for example, to be merely derived intentionality. Rather, it is possible to suggest a naturalistic reduction that accounts straightforwardly for the intentionality of language.[12] From the naturalistic reductionist perspective, the intentionality of the mind and that of language are of the same kind, and the distinction between original (underived) and derived intentionality isn't deep. On Fodor's view,[13] for example, psycho-semantics has priority over lingua-semantics only due to the fact that – or in the sense that – the relevant mind-world causal chains are shorter than the relevant language-world causal chains. In the following discussion I focus on the purported reduction of the intentionality of the mental, but the discussion is supposed to apply to the purported reduction of the intentionality of language as well. Similarly, the criticism to be suggested does not concern any specific naturalistic-reductionist program, but aims at undermining the very idea of intentional naturalistic reductionism.

9 Whereas in Searle 1983. Searle presented a "primitivist" account of intentionality, according to which it is an unanalyzable ground-floor property of mental states, in Searle, 1992, he claimed that intentionality is constituted by the way things appear to the mind, that is, by consciousness.

10 Solipsistic conceptual role accounts such as Block's should be supplemented by a separate external factor for securing hooking up to the world, and thus form dual aspect accounts of semantics.

11 Naturalistic reductionist accounts are offered by, *inter alia*, Block 1986; Dretske 1981; Dretske 1986; Dretske 1988 and Dretske 1995; Fodor 1987 and Fodor 1990; Harman 1982 and Harman 1987; Millikan 1984; Millikan 1989; Millikan 1993; and Millikan 2004; Papineau 1984; Papineau 1987 and Papineau 1993; Stalnaker 1984, and Stampe 1977.

12 Sellars 1963 and Stampe 1977 may be said to suggest such (quite different) reductions.

13 Fodor 1987.

II

What is wrong with the naturalistic reduction of intentionality? It might be instructive to compare the intentional reduction with a scientific reduction, such as the infamous water–H_2O one. What is reduced in this case is watery stuff; that is, stuff with watery properties is given, and it is reduced to something that underlies these properties. We couple properties of two kinds: macro properties (watery properties) and micro properties, and the micro ones are supposed to have explanatory power in relation to the macro ones.

But there is no analogy in this respect between such a reduction and the supposed reduction of intentionality. There are no two properties (or set of properties) in the case of intentionality – there is no analogue to the macro (watery) properties. Let's call the natural relation that is found to be the intentional relation – and that is supposed to be the analogue of the micro property – "R". ("R" can stand for some causal relation that obtains under "normal" conditions, for some evolutionary function, and so forth.) I will now argue that R is not matched with intentionality in its "macro" phase – in a phase that is pre-theoretically given to us.

Let's ask ourselves what the macro analogue with which R is supposed to match is. One may think that the analogue to the macro properties in the case of intentionality is the experience of intentionality, or intentionality as experienced by us. The idea is that our mental states seem to us, from the first person perspective, to be contentful, to have intentional properties, and R is supposed to be matched with these ways our mental states seem to us. One may have doubts regarding the evidential value of introspection as indicator of intentional properties (in contrast, say, to phenomenological properties),[14] but in the present context such doubts are irrelevant, since the present question is whether the ways our experiences and other mental states appear to us – their intentional appearances – may be identified with, or at any rate constrain, the external natural relations. This is supposed to be the point of the micro-macro analogy. And when the question is posed in this way, it becomes clear that the answer to it is negative, for these appearances are not world-involving, and so they cannot be identified with, or explained by, any external natural relations, such as causal ones.[15] If, like Horgan and Tienson[16] or other proponents of "phenomenal

14 I raised such doubts in Horowitz (in preparation).

15 The appearance of intentionality is not to be confused with intentionality of a possibly non-existent (intentional) object – with the "purport to refer" characteristic of mental states (see

intentionality", you take those appearances to indicate the obtaining of intentional properties, then bear in mind that if indeed those appearances indicate any intentionality, they indicate intentionality that is had by the mental *intrinsically*, and intrinsic intentionality imposes no constraints on external relations. So the first-person appearances of intentionality impose no constraints on R; as far as they themselves are concerned, any R goes.

It is important to emphasize that I by no means deny here that there is an appearance of intentionality, or even that this appearance indicates the existence of intentionality. What I deny is that this appearance indicates the existence of something that can be reduced to an external relation.[17] Thus, the claim I am in the process of arguing for is to be understood as denying that there is, in the case of intentionality, a macro analogue mental phenomena that is reducible to an external relation. Had the claim that there is *no* pre-theoretical macro analogue of intentionality whatsoever been established, then first, the continuation of the argument would be redundant, and second, the argument would have established intentional irrealism *unconditionally*, for this claim in itself amounts to the denial of intentionality.

Let's move to consider another candidate for being the macro analogue in the case of intentionality. One might think that intentionality can be identified (in the to-be-reduced level) by means of the syntactical or logical properties of our intentional states. But if such properties are supposed to be what R is matched with, then it must be the case that they endow our intentional states with specific semantic interpretations – that those properties *determine* the semantic interpretations. Yet as Hilary Putnam has shown,[18] logical-syntactical

---

Loar 2003). The latter might be said to be world-involving, in a sense, and thus perhaps to be explainable by a hypothetical external natural relation; the former resists any such explanation.

16 Horgan and Tienson 2004.

17 It might be thought that the appearance of intentionality, being an appearance *of* intentionality, is (like any other appearance, according to this line of thought) intentional. But even if this were true, then since this appearance is not – and it does not indicate the existence of – a mental phenomenon that is essentially world-involving, it cannot serve as the sought macro analogue. Of course, a conclusive case for intentional irrealism (that is not in the scope of this paper) must reject the claim that the appearance of intentionality, as well as any other appearance, is intentional. I believe that the appearance of intentionality should be interpreted in terms of some non-intentional mental reaction on one's part to one's phenomenology.

18 Putnam 1980; Putnam 1981; Putnam 1989.

properties cannot uniquely determine truth conditions.[19] So some external relation *that is not determined by such properties* – some friction with the would-be represented domain – is required for yielding semantic interpretations, but what we are now seeking is not such a relation but a property with which such a relation is to be matched.

A related idea is that intentionality can be identified (in the to-be-reduced level) by means of the functional roles of our intentional states. This seems to be the rationale that underlies various naturalistic-reductionist accounts.[20] The following formal characterization of this idea is offered by Braddon-Mitchell and Jackson: "Belief that p is the theoretically interesting state that actually plays the folk functional roles distinctive of belief that p (from reflection on what we master when we master intentional vocabulary").[21] According to this idea, what this functional role is is a matter of conceptual analysis; R (in my terminology) is that mind-world relation that is in fact possessed by intentional states, and identifying it is an *a posteriori* matter. This suggestion is subject to a difficulty similar to the one that undermines the syntactical-logical identification suggestion: the present suggestion presupposes that the functional roles in question determine semantic interpretations. Do they? We should distinguish here between short-arm (solipsistic) functional roles and long-arm ones. Shoorious for failing to uniquely determine reference and truth conditions – any solipsistic mental characterization is compatible with innumerable semantic characterizaions – and so, trivially , they do not uniquely determine R.[22] As to long-arm functional roles, for their specifications to determine intentionality, it is necessary that they include mention of R itself – of the actual natural intentional relation. That is, the functional role of the intentional state with the con-

19 I am referring to the first part of Putnam's model-theoretic argument, which is quite non-controversial. I do not rely on the whole argument, which is in itself an anti-reductionist argument and is controversial.

20 Papineau 1993, 93 and Papineau 2001.

21 Braddon–Mitchell and Jackson 1997, 481. Braddon–Mitchell and Jackson, as well as Papineau, are concerned with teleological theories of content. Braddon–Mitchell and Jackson argue that such theories cannot conform to this model of reduction (as well as to other ones). Papineau, 2001, who construes the teleosemantic project similarly, responds to their critique.

22 Fodor 1978; Rey 1980; McGinn 1982, and Horowitz 1992. Note that on the suggestion under consideration functional role is supposed to be what determines R rather than R itself – it belongs to the to-be-reduced level rather than to the reducing level. So the suggestion is not a functional role semantic theory of intentionality. But the charge against short-arm functional role theories to the effect that functional roles do not determine intentional objects is effective against this suggestion, in showing that short-arm functional roles do not determine R.

tent that p must include the fact that this state bears R to p. In that case, the identification of R as the intentional relation is a matter of pure conceptual analysis, involving no *a postriori* stage; that is, the present suggestion presupposes that the concept of R is part of the concept of intentionality or reference. I take it that this idea is implausible, but if you are not convinced, I hope the arguments of section III will convince you. We shall then see that the determination of R cannot be a conceptual matter.[23]

Now let us continue to look for a property of intentional states with witch R matches. What other relation (that is, a relation different from R) is (pre-theoretically) given to us? It is not another *natural* relation. It is supposed to be intentionality itself, the intentional relation, as it pre-theoretically appears to us; in other words, it is supposed to be aboutness. But whereas in the case of water we can identify and describe watery properties independently of discovering and describing the microstructure of water, in the case of intentionality we cannot identify intentional properties independently of identifying R. Independently of discovering and describing the natural relation (in some level of abstraction), the only way for us to refer to the phenomenon of intentionality without using the term "intentionality" is to give synonyms or near synonyms to this term, such as "aboutness", "reference", "representation", and "standing for". If asked to point out the properties to be reduced, we will find ourselves engaged with properties in the reducing realm. So we cannot identify any properties that do not belong to the (alleged) reducing level. We can identify those circumstances under which ascriptions of intentionality are made, but in those circumstances we find nothing with which R is matched, for those circumstances themselves are nothing but the obtaining of R.

So the scientific analogy breaks down. Unlike standard scientific reductions, in the naturalistic reduction of intentionality it is not the case that we discover that some pre-theoretically given properties are constituted by some natural relation (the way it was discovered that watery properties are constituted by H_2O). Perhaps we identify an (internal) feature that cannot be subject to reduction to external relations; the external relations that we do identify belong to the reducing realm.

This collapse of the scientific analogy is significant, for the feasibility of the idea of a naturalistic reduction of intentionality in an intentional realistic framework presupposes that there is something that determines that intention-

23 Braddon–Mitchell and Jackson distinguish this suggestion from another one, according to which the natural intentional relation is an essential property of intentional states. This difference makes no difference for my arguments against naturalistic reduction of intentionality.

ality is R. The absence of any macro analogue in the supposed reduction of intentionality – of any relevant "macro" properties – means that nothing constrains R. So R need not match anything in order to be correctly ascribable by content ascriptions – there is no criterion for the truth of the identity statement that the intentional relation is R. Again, any R goes. This entails, in turn, that insofar as some naturalistic relation is supposed to be the intentional relation – insofar as Fodor's dictum is true – there is no criterion for the truth of content ascriptions; more precisely, there is no criterion other than that of correspondence with the practice of content ascriptions. So it can be nothing but this practice itself that determines R as the intentional relation, and sets the standard for what is to count as a true content ascription – for what is to count as a correct application of the concept of intentionality. The truth of an individual content ascription is nothing more than corresponding to the practice of content ascriptions – only relative to such a practice does it have truth conditions and meaning. Thus, content ascriptions do not in themselves have truth conditions. They only have practice-dependent truth conditions, ones that depend on the very practice of these ascriptions (which practice in itself can be said to be neither true nor false). So intentional naturalistic reductionism is intentional irrealism; in other words, if Fodor's dictum is true, then nothing possesses intentionality – nothing *can* possess intentionality.

The line of reasoning suggested here is general, for "R" can stand for whatever natural relation. For example, this line of reasoning undermines the idea of cashing out intentionality in terms of information, i.e., the idea that the content possessed by a representation is the information that it carries (while the notion of information in play is, e.g., the mathematical notion of information or some variation thereof). Since "R" can also stand for carrying information in this sense, then in the absence of "macro" constraint, nothing can make the identification of intentionality and information true (except for the obtaining of correspondence with content ascriptions, which only supports reductionism as intentional irrealism).

Note also, that the case that has just been presented here against the naturalistic-reductionist approach to intentionality – and in turn against intentional realism – does not presuppose that reductions of the intentional to the non-intentional are modeled in any significant way on scientific reductions, so that perhaps another model of such reductions would avoid the critique under consideration. I argued that in the absence of properties analogous to the macro ones, nothing constrains the identity of the supposed intentional natural relation except for the practice of content ascriptions. Such a constraint is required for whatever reduction of intentionality, and so this lack of constraints means that the very idea of intentional naturalistic reductionism entails intentional irrealism. The only similarity to the scientific reduction that is assumed in the argument is that some (macro-analogue) property is supposed to constrain the reduction, and the mere unavailability of such a property is what leads to the failure of the idea of the naturalistic reduction of intentionality according to this argument.

It is important to emphasize that it is the semantic conception of intentional irrealism that is conditionally vindicated: the point is not that some, or many, or most, content ascriptions are false, if Fodor's dictum is true – the argument has nothing to do with the actual truth value of content ascriptions; rather, the point is that, in case Fodor's dictum is true, all content ascriptions do not, in themselves, have truth conditions. This is a strong claim, which might appear to fly in the face of reason. For it seems that we know the truth conditions of content ascriptions, that we regularly recognize them, and so that, *a fortiori*, even if we are badly mistaken about the truth values of content ascriptions (for failing to recognize whether they obtain or not), they do have truth conditions.

We may question the plausibility of the semantic conception of intentional irrealism from another direction, having to do with the concept of intentionality: it seems that we regularly apply this concept – or, if you will, the concepts of reference, think about, etc. So this concept appears to have application conditions, which means that the semantic conception of intentional realism appears to be true. Perhaps, then, if we take Fodor's dictum for granted, we should look for an alternative understanding of the notion of the reduction of the intentional, one that avoids the irrealistic implication I attributed to this notion.

III

Indeed, I shall now examine an alternative understanding of the notion of the reduction of the intentional that might be taken to be underlain by the idea that the concept of intentionality (or of reference, or of think about, etc.) has application conditions. Some philosophers hold that the connection between intentionality and the natural relation in question is conceptual; that is, it is *the concept* of intentionality (or of reference, or of aboutness, etc.) that determines R as the intentional relation; that R is what this concept is a concept of. Thus, in explaining what makes one – among the many word-world relations – the relation of reference, Michael Devitt writes:

> Why is the relation between Bill Clinton and Chelsie Clinton, Prince Philip and Prince Charles, and [...] rather than the relation between Bill Clinton and Newt Gingrich, Prince Philip and Baroness Thatcher [...] to be identified as *being the father of*? Or the analogous question about the property *cathood*: Why is the property of Nana, Jemina, and so on [...] rather than the property of Fido, Bill Clinton, and [...] to be identified as *cathood*? We need say no more in answer to these questions than that the specified relation just *is being the father of* and the specified property just *is cathood*. Similarly, we need say no more in answer to the question about reference than that the specified relation just *is* reference. That relation and no other simply is the one we call "reference" and that's that.[24]

The idea is that the identity of the natural relation that is supposed to constitute intentionality is a conceptual matter. The reference relation, or the intentional relation, is simply that one among the many (language-world or mind-world) natural relations that obtain in the world to which concepts such as "refers to" or "thinks that" apply. So it is this concept which determines R and connects it with intentionality: it is this concept with which R is supposed to match. (We should not interpret this suggestion as committed to there being an explicit definition of "intentionality" or "reference". Plausibly, the idea is that R is "chosen" by the use of ascriptions of content and reference in various circumstances.)

This idea seems to make sense. We regularly apply the concept of intentionality (or the concepts of reference, or that of think about, etc.), and when we apply it from the third-person perspective, then, plausibly, its application conditions are "naturalistic".[25] The semantic conception of intentional realism is thus affirmed as a form of intentional naturalistic reductionism, that gets a clear sense. Of course, this is not the story of scientific reduction – this is the standard story of concept application. But the question that we have to face is whether this story can sustain a realistic interpretation of intentional reductionism.

The water–H_2O analogue of the present suggestion would be the idea that H_2O is simply (identified as) what we label "water". That means that the watery stuff, the watery properties, are left out of the picture; they play no essential role in the truth conditions of the identity statement "Water is H_2O". This is certainly implausible, but this implausibility does not in itself constitute a reason against the suggestion in question, since this suggestion need not be taken to apply to the H_2O case and similar cases. So let's examine whether the suggestion itself – as an account of the natural intentional relation – is plausible, or

---

24 Devitt 1997, 332–333.

25 The alleged holistic character, on the one hand, and normative character, on the other, of content ascription might be taken to undermine the claim that the truth conditions of third-person content ascriptions are naturalistic. I think proponents of this claim can rebuff these charges, but none of the theses for which I argue in this paper depends on this.

whether, specifically, it founders on the same rocks as that H_2O–water conceptualist identification.

It seems to me that the latter alternative prevails; that this conceptualist account of the natural intentional relation also leaves out something that it shouldn't. It seems that aboutness is left out of the picture. Consider. The statement that identifies intentionality with some natural relation is supposed to be about a certain phenomenon that is pre-theoretically given to us – about aboutness; as such, it must be a synthetic statement, and not merely one of labelling some relation – not merely a stipulation. If the naturalistic-reductionist approach is a realistic approach to intentionality – if it takes intentionality to be real and aims to inform us of its real nature – then it is committed to the idea that we first encounter the phenomenon of intentionality, namely aboutness, and then identify that natural relation that underlies this phenomenon (it's not that we learn what intentionality is upon encountering R). This means that the identity statement in question cannot be a stipulation of the sort under consideration. "Intentionality" is not supposed to be just a name for some natural relation, a name that could be replaced (in statements such as "This is a case of intentionality" which accompany demonstrations of R) with any arbitrary combination of letters. And this is, in fact, what the present suggestion delivers – mere labelling that has nothing to do with aboutness. Trivially, such a labelling affirms the obtaining of R, but our issue is that of intentional realism, not that of R-realism; intentionality, or the concept of intentionality, plays no role in stipulating that a certain term – be it "intentionality", "reference", or whatever other term – denotes R.[26]

---

26 One might argue that what underlies the fact that it is R that underlies our content ascriptions – what underlies the fact that intentionality refers to R – is the fact that citing R in explanations and predictions of behavior is most efficient; that is, R is that relation appealing to which explains behavior better than appealing to any other relation. (Devitt himself has a suggestion along these lines.) I wonder whether it is possible to justify the identification of the intentional relation with the single most efficient natural relation in this respect (i.e., whether such an identification matches our practice of content ascriptions), but suppose it is. The proponent of this suggestion still has to say what connects explanatory efficiency with intentionality, and the suggestion seems to fall into the same predicament described above. Note the difference between the suggestion in question and intentional instrumentalism such as Dennett's 1978; Dennett 1987: on intentional instrumentalism R is the relation of explaining and predicting behavior. On the present suggestion R is some natural relation (e.g., some causal relation) whose citing enables the explanation and prediction of behavior – R is a different relation than that of explaining and predicting behavior.

As noted, the problem that Devitt's conceptualist notion addresses is that of there being multitude of natural word-world – and, similarly, mind-world – relations: what determines one of them as the intentional relation? The macro-micro analogy can be taken to address this very question. According to it, it is the (pre-theoretically given) phenomenon of aboutness that constrains those relations. I argued that this reply founders on the rocks of the impossibility to locate relevant macro-like phenomena in the case of intentionality (such that can be accounted for by means of external relations). Devitt's alternative reply is in fact that it is a stipulation (Devitt of course does not use this term) that "chooses" the relevant natural relation, and I argued that this reply in fact leaves out intentionality. So the conceptualist reduction of intentionality fails to connect the supposed intentional phenomenon with a natural relation via the concept of intentionality, and it cannot rescue this phenomenon: it will not do to say that some R is the relation called "reference", or that this is how we ascribe content. This leaves us with no external criterion for the truth of content ascriptions, that is, with intentional irrealism. (Again, this argument too applies to any natural relation whatsoever; one may call information, or indication, or whatever other natural relation, "intentionality" – any such analytic identification would leave intentionality out.)

But further, the very idea of connecting intentionality with R via its concept is ill-founded. The assumption on which this approach is based is false, for the concept of R (be R *whatever* natural relation) is simply not the concept of intentionality, or of reference, or of think about, etc. For the concept of intentionality, that is, the concept of *aboutness*, has nothing to do with any specific natural relation. It is neutral with respect to the identity of the natural relation, and does not prefer any specific relation over any other. Again, taking the connection between intentionality and R to be conceptual in the way under consideration is an empty stipulation that robs "intentionality" of any content, and thus of any content we take it to have.

Furthermore, had the concept of intentionality been the concept of a specific natural relation, then any two philosophers who differ in what they take to be their theories of intentionality (that is, who differ in their views about the identity of the natural relation to be associated with "intentionality") would not have been able to understand each other when speaking about the referent of "intentionality" – in fact they would not have been referring by "intentionality" (and "reference" etc.) to the same phenomenon. This also holds for any two people who disagree about the content of any entity without having disagreements about this entity's non–semantic properties and relations.

But one might think that, nevertheless, we can adhere to the claim that the concept of intentionality (the concept of aboutness) is the concept of R. Suppose that we adopt the externalist idea that one's macro concepts (e.g., the concept of water) are the concepts of whatever micro properties (e.g., H_2O) that in fact – whether or not one knows it – underlies the relevant macro properties, and apply this approach to the relations between the pre-theoretic reduced concept of intentionality and R.[27] According to this suggestion, one's concept of intentionality is the concept of whatever external relation that in fact underlies ascriptions of intentionality. It may thus be the concept of R, even though no pure *a priori* conceptual analysis can reveal R in it.

I take this externalist idea to be flawed,[28] but I can grant it in the present discussion since it cannot rescue intentional naturalistic reductionism from the irrealistic fate. For on such an approach, it is the connection between the reduced and reducing properties – e.g., watery properties on the one hand and H_2O on the other – that is supposed to be responsible for the macro concept being the concept of the specific micro property in question. Thus, the truth of the claim that the concept of intentionality is the concept of R depends on the supposed connection between intentionality in its macro phase and R. But since no "macro" properties are available in the case of the alleged intentional reduction, then – unlike the water–H_2O identification – there is no possible ground for this identification, no criterion for its truth. Note how the externalist characterization of the concept of intentionality under the current suggestion betrays its irrealistic nature: this concept is supposed to be the concept of whatever external relation that in fact underlies ascriptions of intentionality. That is, R can only be constrained by content ascriptions. This means that to the extent that intentional realism depends on the conceptualist reduction construed externalistically, there is no criterion that is external to the practice of content ascriptions for the truth of such ascriptions. So the truth of any content ascription is practice-dependent, and so intentional realism is false. The naturalist-reductionist finds herself in the same predicament.

---

27 See Putnam 1975. It is legitimate to assume externalism in criticizing my argument even if externalism is understood as a realistic thesis, since the burden of proof is with me to show that intentional realism (of all varieties) is false within the naturalistic reductionist framework. Note also, that externalism can be construed as a thesis regarding content ascriptions (a thesis to the effect that such ascriptions are sensitive to external factors) and thus as non-committal with respect to the issue of intentional realism versus intentional irrealism.

28 See Horowitz, 2001 and Horowitz, 2005. In these papers I took the issue to concern the nature of content rather than content ascriptions.

IV

If the arguments presented here are along the right lines, then the very idea of the naturalistic reduction of intentionality – when realistically interpreted – is flawed. As noted, we have arrived at a strong conclusion, namely that, if Fodor's dictum is true, then all content ascriptions lack (practice-independent) truth conditions; it is the falsity of the semantic conception of intentional realism that has been shown to follow from Fodor's dictum. The conceptualist approach to intentional naturalistic reductionism appeared to promise a straightforward vindication of this conception, but we see that it cannot deliver what it promised. The truth-conditions that content ascriptions appear to possess are not possessed by these ascriptions in themselves, but are practice-dependent – they depend on and are endowed by the practice of content ascriptions. Similarly, the application conditions of the concept of intentionality (or of reference, or of think about, etc.) are "determined" (in a qualified sense – see below) by the practice of concept ascriptions. This means that in the intentional naturalistic-reductionist picture, the concept of intentionality does not determine its own conditions of application.

One may wonder how there could be a concept that does not determine its own conditions of application. This wonder should dissolve, however, once we recall that to reject the semantic conception of intentional realism is to reject the idea that the concept of intentionality could be a concept of a real property. I argued above that the conceptualist approach to intentional naturalistic reductionism robs the concept of intentionality, the concept of aboutness, of any content we take it to have. What is this content on my approach? The problematic character of this concept makes it impossible to articulate an answer to this question. There are ample examples in the history of philosophy of concepts (or concepts*) that seem to us to be perfectly lucid and coherent, but that, upon philosophical reflection, are revealed to lack a clear sense and integrity, and I submit that the concept of intentionality (or that of reference, or think about, etc.) is such a concept.

It should be clarified that the claim that content ascriptions have practice-dependent truth conditions does not mean that entities such as beliefs may have intentional properties that are practice-relative or relational (that they may have them relative to the practice) in the way in which secondary qualities are supposed to be relational properties of physical objects, or, similarly, in the way in which semantic externalism takes intentional properties of beliefs (narrowly considered) to be environment-relative. Both kinds of cases allow some (relational) form of realism with respect to the properties in question, but the practice-dependence view of intentionality that I suggest is significantly different, since according to it *the truth values of content ascriptions depend on nothing but the truth values of content ascriptions*, and there is no external criterion for the truth values of any content ascription.

Here is another way to show this point: the obtaining of a practice of content ascription merely determines what contents are sanctioned by the practice that in fact obtains; it does not determine that entities in fact have the contents sanctioned by it; it does not and cannot determine that it itself is *binding*, and thus it does not endow any entity with content. In this respect, the obtaining practice has no advantage – as far as determining the *truth values* of content ascriptions are concerned – over any imaginary practice. (We should similarly understand the role of the practice in "determining" the application conditions of the concept of intentionality.) So, again, there is no external criterion for the truth value of any content ascription in the naturalistic-reductionist framework, and this means that, in this framework, no content ascription in itself has truth-conditions; put differently, no concrete entity has any intentional properties.

So if intentionality has to be something else, it isn't real. Since, as noted, I accept Fodor's dictum, namely that if intentionality is real it must be something else, I endorse the conclusion that intentionality isn't real.

This is a radical conclusion. Since the claim that content ascriptions lack absolute truth-conditions means that nothing can have intentional properties, it seems to follow that nothing – e.g. no sentence, no belief – has truth-conditions. It might then seem that we cannot distinguish true beliefs from false ones – that all beliefs have the same epistemic status (if any). This might seem troubling. There seems to be a good reason to maintain that there are epistemic differences among beliefs. However, repudiating intentionality means the absence of truth-conditions – and thus of the ability to distinguish among beliefs in terms of epistemic status – in a realistic-representational sense. It does not rule out the possibility that beliefs are true or false in an alternative sense, if there is any alternative sense. I believe that they are true and false in the pragmatist sense.[29] Specifically, I endorse the main tenets of Quinean holistic pragmatism. Of course, defending this picture, and the way intentional irrealism of the genre I endorse can find its place in it,[30] should await another occasion.

---

29 The pragmatism in question is pragmatism with respect to truth, and it should not be confused with pragmatism with respect to meaning, such as the verificationist approach to meaning that was advocated by Peirce (1878), 1992.

30 Since on the Quinean view nothing compels us to either accept or reject any claim, one who wishes to defend a unique irrealism of the intentional should be able to explain the sense in which content ascriptions differ in epistemic status from standard (non-semantic) claims.

Let me make two final remarks. First, the irrealistic conclusion we've arrived at allows for a reductionist-naturalistic theory to be true when abstracted from realistic aspirations (from accounting for how thought or language hooks up to the world); it allows it to be regarded as a theory that systemizes the regularities governing our use of semantic idioms such as "reference", "think of", etc., and that's that. On this approach, such a theory is a theory of our practice of (third-person based) content ascriptions, and of nothing more – a theory of ascriptions of intentionality, rather than of intentionality. Of course, the truth of any such theory, thus understood, presupposes that the conditions that underlie our (third-person based) content ascriptions are naturalistic. I believe that this latter claim is true, and I believe that naturalistic theories can be enlightening, and some of them are. Still, I doubt that we discovered, or that we will ever discover, a reductionist-naturalistic theory that correctly identifies the natural relation that underlies our practice of content ascriptions. For the human practice of content ascriptions is extremely varied and complicated – it is too varied and complicated to be capturable by any single crispy formula – and plausibly counterexamples will always be awaiting around the corner.

Second, my repudiation of intentionality is not a repudiation of beliefs, desires, and the other so-called propositional attitudes. Stich's eliminativism[31] was based on the thought that since no entity possesses all the properties that are supposed to be possessed by beliefs, then there are no beliefs. Later Stich[32] realized that this reasoning presupposed a specific theory of reference, and adopts a pragmatist approach to the question of whether there are beliefs. I will not dwell on this issue. I only wish to emphasize that I am not committed to denying the existence of beliefs and the other so-called propositional attitudes in this sense: everything I said allows for there being belief states *minus* intentional-semantic properties. That is, it allows for there being states with the phenomenological properties (and among them phenomenology of intentionality), epistemic properties, syntactic properties, and causal-functional properties of beliefs, but that lack semantic properties. Of course, for many philosophers this – to use Fodor's apocalyptic expression – is the end of the world. I believe that it isn't, but showing this will also have to await another occasion.

31 See Stich 1983, and also Stich 1978.
32 Stich 1996.

References

Block, N., 1986, "Advertisement for a Semantics for Psychology". *Midwest Studies in Philosophy*, 10, 615–678.
Boghossian, P., 2008, "Is Meaning Normative?" In: *Content and Justification*. Oxford, Oxford University Press.
Braddon-Mitchell, D. and Jackson, F., 1997, "The teleological theory of content". *Australasian Journal of Philosophy*, 75, 474–489.
Brandom, R., 1994, *Making it Explicit*. Cambridge (Mass.), Harvard University Press.
Dennett, D.C., 1978, *Brainstorms*. Cambridge (Mass.), MIT Press.
Dennett, D.C., 1987, *The Intentional Stance*. Cambridge (Mass.), MIT Press.
Devitt, M., 1997, *Realism and Truth* (2nd edition, with a new afterward). Princeton, Princeton University Press.
Dretske, F.I., 1981, *Knowledge and the Flow of Information*. Cambridge, Mass., MIT Press.
Dretske, F.I., 1986, "Misrepresentation,", in R. Bogdan (ed.), *Belief: Form, Content and Function*. Oxford, Oxford University Press.
Dretske, F.I., 1988, *Explaining Behavior*. Cambridge (Mass.), MIT Pres.
Dretske, F.I., 1995, *Naturalizing the Mind*, Cambridge (Mass.), MIT Press.
Field, H., 1978, "Mental representation", *Erkenntnis*, 13, 9–61.
Fodor, J.A., "Tom Swift and his Procedural Grandmother", *Cognition*, 6, 229–247.
Fodor, J.A., 1987, *Psychosemantics*. Cambridge (Mass.), MIT Press.
Fodor, J.A., 1990, *A Theory of Content and Other Essays*. Cambridge (Mass.), MIT Press.
Gibbard, A., 2012, *Meaning and Normativity*. Oxford, Oxford University Preess.
Harrman, G., 1982, "Conceptual Role Semantics", *Notre Dame Journal of Formal Logic*, 28, 242–256.
Harrman, G., 1987, "(Nonsolipsistic) Conceptual Role Semantics", in: E. LePore (ed.), *New Directions in Semantics*. London, Academic Press.
Horgan, T. and Tienson, J., 2002, "The Intentionality of Phenomenology and the Phenomenology of Intentionality", in: D. Chalmers (ed.), *Philosophy of Mind: Classical and Contemporary Readings*. Oxford, Oxford University Press.
Horowitz, A., 1992, "Functional Role and Intentionality", *Theoria*, 58, 197-218.
Horowitz, A., 2001, "Contents just are in the Head", *Erkenntnis*, 54, 321–344.
Horowitz, A., 2005, "Externalism, the Environment, and Thought-tokens", *Erkenntnis*, 63, 133–138.
Horowitz, A. (in preparation), "Conscious Intentionality and its Alleged First-Person Perspective".
Husserl, E., 1913, *Logische Untersuchungen*. Halle, Niemeyer.
Kripke, S., 1982, Wittgenstein on Rules and Private Language. Oxford, Blackwell.
Loar, B., 2003, "Phenomenal Intentionality as the Basis of Mental Content", in: Hahn, M. and Ramberg, B. (eds.), *Reflections and Replies: Essays on the Philosophy of Tyler Burge*. Cambridge (Mass.), MIT Press.
McGinn C., 1982, "The Structure of Content", in: Woodfield, A. (ed.), *Thought and Object: Essays on Intentionality*. Oxford, Clarendon Press.
Millikan, R.G., 1984, *Language, Thought and Other Biological Categories*. Cambridge (Mass.), MIT Press.

Millikan, R.G., 1993, *White Queen Psychology and Other Essays for Alice.* Cambridge (Mass.), MIT Press.

Millikan, R G., 1989, "Biosemantics", *Journal of Philosophy*, 86, 281–297.

Millikan, R.G., 2004, *Varieties of Meaning: The 2002 Jean Nicod Lectures.* Cambridge (Mass.), MIT Press.

Papineau, D., 1984, "Representation and Explanation", *Philosophy of Science*, 51, 550–572.

Papineau, D., 1987, *Reality and Representation.* Oxford, Blackwell.

Papineau, D., 1993, *Philosophical Naturalism.* Oxford, Blackwell.

Papineau, D., 2001, "The Status of Teleosemantic, or How Stop Worrying about Swampman", *Australasian Journal of Philosophy*, 79, 279–289.

Peirce, C.S. 1878/1992, "How To Make Our Ideas Clear", in: Houser, N. and Kloesel, C. (eds.) *The Essential Peirce*, Vol.1. Bloomington, Indiana University Press.

Putnam, H., 1975, *Mind, Language, Reality: Philosophical Papers Vol 2.* Cambridge, Cambridge University Press.

Putnam, H., 1980, "Models and Reality", *Journal of Symbolic Logic*, 45, 464-482.

Putnam, H., 1981, *Reason, Truth and History.* Cambridge, Cambridge University Press.

Putnam, H., 1989, "Model Theory and the 'Factuality' of Semantics", in: George, A. (ed.), *Reflections on Chomsky.* Oxford, Blackwell.

Quine, W.V O., 1960: *Word and Object.* Cambridge (Mass.), MIT Press.

Rey, G., 1980, "The Formal and the Opaque". *Behavioral and Brain Science*, 3, 90–92.

Searle, J.R., 1983, *Intentionality*: An Essay in the Philosophy of Mind. Cambridge, Cambridge University Press.

Searle, J.R., 1992, *The Rediscovery of the Mind.* Cambridge (Mass.), MIT Press.

Sellars, W., 1963, *Science, Perception, and Reality.* London, Routledge.

Stalnaker, R., 1984, *Inquiry.* Cambridge (Mass.), MIT Press.

Stampe, D. W., 1977, "Towards a Causal Theory of Linguistic Representation", *Midwest Studies in Philosophy*, 2, 42–63.

Stich, S., 1978, "Autonomous Psychology and the Belief-Desire Thesis", *The Monist*, 61, 571–591.

Stich, S., 1983, *From Folk Psychology to Cognitive Science.* Cambridge (Mass.), MIT Press.

Stich, S., 1996, *Deconstructing the Mind.* New York, Oxford University Press.

Bernhard Obsieger

Husserlian Ethics

The Morality of Action

1 Introduction

Originally, following Brentano and Kant, Husserl assumed that ethics was principally concerned with the universal moral laws that are valid for actions. He accordingly believed that ethics must investigate the grounds and criteria for the moral correctness of actions and of their motivation. In light of this problem, Husserl establishes a parallelism between what is right in a logical and in a moral sense. There is always one action that is morally right, namely, that which is the *best* under the given circumstances. What makes an action better than another is that it realizes a higher good, and the best action available is therefore the one that realizes the highest available good, that is, the one of the *highest value*. This general criterion of ethical action, requiring that we always strive to realize the highest possible value (which can eventually consist in a sum or collection of different but compatible values), is what Husserl calls the categorical imperative. Despite the Kantian term, he takes this principle as such from Brentano, who proposed it as an alternative to that which Kant himself considers the supreme principle of moral action.[1]

Husserl agrees with Brentano that the ultimate source of moral knowledge must be understood as a peculiar kind of experience, namely, as the realm of positive and negative emotions that Brentano calls acts of love and hate, to which also belong the relational acts of preferring. As Brentano points out, emotions of love, hate, and preference can be intrinsically characterized as right or wrong, and thereby they are a source of experiential knowledge – precisely the moral knowledge of what is good or bad, better or worse. That which is loved or preferred in acts that are given as "right" is thereby experienced as good or as better than something else; that which is hated in negative emotions that are characterized as "right" is bad. Conversely, the objects of those positive emotions or acts of love that are characterized as wrong are bad, and the objects that

1 Husserl 1988, 43, 129 ff., 237 ff. (On Husserl's attitude toward Kant's version of the imperative, cf.: 414 ff.) Brentano 1955, 36–40, and Brentano 1952, § 13.

https://doi.org/10.1515/9783110560282-165

are wrongly preferred are not better but worse than the objects to which they are preferred. The reflection upon the intrinsic rightness or wrongness of the emotions in which we take a positive or negative stance toward certain objects is thus the source of our knowledge of the goodness or badness of these objects, a knowledge that according to Brentano is the foundation of morality.

Although Husserl agrees with Brentano that the evident rightness or wrongness of emotional experience is the source of moral knowledge, he does not understand this knowledge as the inner consciousness of subjective *acts* that are mysteriously characterized as right or wrong.[2] Rather, he argues that these acts are a source of knowledge and can be right or wrong because they give us intentional and intuitive access to a certain realm of *objects* or qualities of objects, namely, to the realm of values. Values are those practical properties that make something *worthy* of being loved or hated or *more preferable* than something else, and thus are responsible for the rightness or wrongness of the subjective acts of love or preference. It is by way of these acts that values are present in our experience, and if our emotions are characterized as immediately motivated by the values of their objects and as corresponding to these values, they are characterized as right. Conversely, if they are characterized as not corresponding to the values and their relative ranking, this means that they are intrinsically characterized as wrong. In accordance with other disciples of Brentano such as Meinong and Ehrenfels (and also with Scheler), Husserl thus conceives moral knowledge not principally from its subjective side (as concerning the correctness of the *acts* of love and hate), but first and foremost in light of the values that are the objects of this knowledge.[3]

Insofar as Husserl adopts Brentano's version of a categorical imperative, he initially understands the relation between the formal and the material dimension of moral action in basically the same way as Brentano, but he gives much more weight to the formal dimension. Brentano already formulated a set of general laws concerning the sphere of preference,[4] and he called attention to the formal laws that determine the moral rightness or wrongness of action. According to Brentano, moral action is guided by a general attitude that is character-

---

2 Husserl 1988, 344. As J. M. Palacios nicely points out, Brentano reached a point where he would have been forced to introduce the notion of values—thereby abandoning his one-sided, subjectivistic approach to moral knowledge and to intentionality in general—but he declined even to ask the obvious questions concerning what it is that makes preferences right or wrong or what the objects (or objective correlates) are to which preferences refer. Palacios 2008, 31–44, especially, 40–42.

3 Husserl 1988, 276 f.

4 Brentano 1952, 211 ff.

ized by a *formal* principle. Husserl takes this principle as a point of departure for the development of a system of formal ethical laws, and he explores and formalizes the general relations between goods and acts of love and preference that were discovered by Brentano. This formalistic approach is closely connected with Husserl's parallelization of ethics and logic. Once again it was Brentano who first pointed out this parallelism between the sphere of logic and that of morality by establishing a close analogy between truth and goodness or between the rightness of judgment and the rightness of the moral emotions of love, hate, and preference.[5] But it was left to Husserl to explore this formal dimension of ethical knowledge systematically and to develop concrete foundations of a formal practical discipline analogous to formal logic.

Despite Husserl's interest in the formal structure of morality, which is probably influenced by the Kantian formalistic understanding of ethics, Husserl reserves a decisive role for the *material* dimension of values. After all, the formal laws always depend upon the material knowledge of which values are ranked higher or lower than others, or as Brentano would put it, which preferences are characterized as right or wrong.[6] As in other areas of knowledge, the knowledge of moral demands does not primarily consist in the knowledge of formal ethical laws, because the applicability of these laws depends upon the material knowledge of the realm of values as the matter to which the formal laws refer.[7] Furthermore, the universal material knowledge of values and their relations must be complemented by the empirical knowledge of the concrete values that are at stake in a given situation. Insofar as all values and their ranking belong to the matter of ethical knowledge, it is the matter that principally determines what is morally correct. What is more, the value of something depends on what and how it is, and therefore the knowledge of values depends in turn on another

---

5 Brentano 1952, 134 ff. Brentano admits that there is no equivalent to preference (and the relation of "better" or "worse") in the sphere of judgment, but he nevertheless affirms an analogy between the intrinsic rightness of preference and the evidence that characterizes judgment as true (Brentano 1952, 147 f.). Husserl himself explicitly acknowledges that it was Brentano who inspired his formalistic approach to ethics and was the first to formulate formal axiological laws, even though Brentano's formulation referred only to the psychological or subjective dimension. Husserl 1988, 90.

6 Husserl 1988, 140.

7 This reference to a restricted material realm rather than to "something in general" thus introduces an important difference between the meaning of ethical and logical formality. Formal ethical laws do have a certain material content insofar as they express the nature of a particular sphere of reality, namely, that of volition. Husserl 1988, 67 f. and 139.

kind of material knowledge, namely, the knowledge of the objects that possess them.

For these reasons, formal ethical knowledge must be based in several ways upon the material knowledge of values and valuable objects. For instance, it is certainly a formal law that choosing the realization of the highest value available is the correct choice. It is also a formal law that it is better to realize a higher value than a lower one, or that it is better to realize the lesser evil when confronted with a choice between negative values, and so on.[8] However, *what* kinds of things have *which* positive or negative values cannot be addressed in such a formal manner, but can only be subject to material knowledge. Consequently, Husserl compares the relation between the form and the matter of ethical laws to the relation between the form and the matter of theoretical knowledge and ascribes the decisive function to the side of the matter.[9]

Thus in the same way as Scheler and even earlier than him, Husserl considers the morality of action as depending upon the concordance of our preferences and choices with the order of values.[10] Which action is good or bad depends upon the right choice, which depends in turn upon the rightness of our preferences, that is, of their correspondence with the order of values. Husserl emphasizes the formal laws that govern the relation between what Scheler later called the subjective and the objective orders of values (or orders of love).[11] There is a form of right and wrong action, and this form consists in the relation between the order in which values are in fact preferred and chosen, on the one hand, and their objective order on the other hand. Apart from his interest in the formal dimension of ethics and the thesis of a parallelism between ethics and logic, Husserl's position in the years between 1908 and 1914 thus seems very close to the phenomenological ethics of Max Scheler. Nevertheless, unlike Scheler, Husserl conceives the relation between axiological acts (love, hate, preference) and the realm of values not as an adequation (or inadequation) between two separate orders—one concerning how we rank the values and the other concerning their objective ranking—but as a correspondence between our assessment of values and their intuitive presence in our emotional experience. By appealing to the Brentanian distinction between evident and non-evident acts of love and preference, Husserl thus avoids Scheler's thesis that morally

---

8 For this kind of formal ethical laws, Husserl 1988, 92 ff.

9 Husserl 2004, 235.

10 See, for example, Husserl 1988, pp. 90 ff. and p. 140. At least the first of these texts already belongs to the first version of these lectures from 1908/09; cf. the editor's introduction, xiv.

11 Scheler 2000, 348 ff.

wrong choices are rooted in a mistake concerning the objective order of values, which leads Scheler to renew the Socratic interpretation of evil as a form of error. According to Husserl, the subjective realm of preferences is not the source of a private order of values that correctly or incorrectly mirrors their real order, producing a subjective "order of values *according to me*" that can be opposed in Scheler's manner to an objective order of values that subsists independently of any experience. Rather, the objective order of values is the correlate of those acts of love and preference that are characterized as an intuitive self-presence of the corresponding values. Our assessment of values can only be adequate or inadequate in virtue of its relation to the presence of values in our very acts of ranking them, but not all such acts possess the same evidence and thus the same authority. Consequently, Husserl avoids Scheler's dichotomy of a subjective and an objective order of values by applying Brentano's insight that our axiological *experience* can be characterized as right or wrong in itself, although this is only because this experience is not a separate and self-sufficient realm, but consists in the intuitive presence of its object.

This conception of axiological experience allows Husserl to avoid the problems connected with Scheler's objectivist and Platonist ontology of values. Unlike Scheler but like Brentano, Husserl thus sees the foundation of moral rightness not in an external correspondence between our emotive experience and its object, but rather in our experience itself, avoiding at the same time Brentano's subjectivist reduction of values to the acts in which they are given. In this way Husserl's account of moral rightness is strictly phenomenological and avoids Scheler's ontological commitment to the mind-independent existence of an ideal order of values.

However, the 1908–1914 lectures still do not contain Husserl's most original contribution to ethics. At this period Husserl still addresses ethics in the traditional manner as a separate and rather secondary philosophical "discipline," and he still considers the realm of moral demands as a universal objective sphere that does not concern the agent's personal and vital situation. But in his later work, Husserl comes to question this understanding of morality, rediscovering a dimension of ethics that used to be of central importance in antiquity, namely, that of an ethical *life*. This leads him to understand ethics primarily as a guide to a happy and valuable life that encompasses all kinds of human activity, including the acquisition of theoretical knowledge. Insofar as all human knowledge ultimately serves the purpose of a genuinely human life in its interaction with the world, ethics can no longer be understood as one particular philosophical discipline among others, but must provide the foundations for human knowledge in its entirety.

2 The relativity of moral obligations to the history of each individual life

Brentano's categorical imperative basically consists in the requirement of always choosing the objectively best option, that is, the option that one believes to be the best after comparing the values of the different possibilities at hand. This presupposes that the preferable option is one and the same for everyone and that *who* is confronted with the possibilities in question is not ethically relevant. Yet such a limitation to objective values seems to be incompatible with our basic moral beliefs. As Husserl argues, a mother has obligations toward *her* child that she does not have toward any other child. If she has to choose between saving the life of her child or the life of someone else, it would be wrong to say that she should ask herself which of these lives promises to be objectively of the highest value.[12] What is "objectively" the same choice is not the same for different persons, and their duties can be different although they are faced with the same possibilities and the same objective values. Husserl therefore says that each person has her own categorical imperative that is different from the categorical imperative of everybody else.[13]

This critique shows that the duties of a person have to do with her own situation in life and her personal obligations toward other persons. What is demanded in a given situation largely depends on the history of one's life, on the compromises we have made with others and ourselves, as well as on the decisions we have previously made and the projects in which we are engaged. We never experience situations in an objective and isolated manner, as situations that are what they are in themselves and for everyone. Rather, one experiences each situation as a part of one's life and in its relation to one's life as a whole. It is thus an abstraction to consider situations and actions isolated from the context of a particular life, even though this abstraction is to some extent legitimate and necessary, since there are many situations that concern universal structures of social life, such as professional or family life, in which we are involved not as an individual person, but as someone endowed with a social role. Just as different persons can have different possibilities, so also their duties are not always the same even if they are faced with one and the same objective possibility. To use the famous example from Plato's *Euthyphro*, it is not the same whether someone you accuse of a crime is unknown to you or is your father. Since a

12 Husserl 2014, 390–392.

13 Husserl 2014, 391.

social identity is inseparable from the context of a human life, then even in the case of universal and standardized situations, one's choices and actions are ultimately only understandable as part of the history of one's life as a whole. This life is at the same time an individual life and a social life and is in itself historical in these two different ways, as an individual historical process and as taking part in the history of different communities and in the universal history of humanity.

Accordingly, moral responsibility is not primarily the responsibility for this or that action, but for our life as a whole. The practical sphere as the central topic of ethics must not be limited to isolated situations, but should be understood as extended to the most basic and all-encompassing responsibility that concerns the question of what we should do with our lives. Although each person can only freely dispose of her own life, the structure of "living one's own life" is a universal structure that can and must be addressed by ethics. Because all particular actions are ultimately only understandable as parts of the all-encompassing action of living one's life, human life as a whole becomes the central topic of Husserl's ethical reflections. This focus on life thus replaces the traditional focus on particular actions and characterizes Husserl's later manuscripts on ethics published in *Die Grenzprobleme der Phänomenologie*.

3 The Structure of Ethical Life

Our more concrete decisions are normally taken on the ground of decisions of a higher order, in such a way that the decisions concerning the means are founded upon the decisions concerning the ends. What is ultimately an end and gives meaning to our actions is that which is chosen for its own sake, because it is taken to be good or desirable in itself. The possibilities that shape our life and are desirable or undesirable in themselves – and thus endowed with positive or negative value – are of two different kinds. On the one hand, there are possibilities that concern the way in which we live the present moment or another particular moment that is still to come, and on the other hand, there are possibilities of permanent value that relate to one's life as a totality. The latter possibilities concern the form and the overall meaning of one's life. Each of these two kinds of possibilities can be either of positive or of negative value.

An important share of the decisions concerning one's life as a whole is related to the way in which we want to participate in the life of our community. There can be many possible occupations that attract us, and choosing those to which you want to devote your life concerns the dimension of human existence that Husserl calls professional or vocational life.[14] This kind of life emerges out of a free choice (of course dependent on our circumstances) that is the source of obligations and duties. Vocational life encompasses both professional life in the narrow sense and family life or even social life in general, and it is that which gives its structure to every human life.[15] Although vocational and professional life is based upon one's preference for certain values at the cost of others, our vocational preferences are prior to voluntary choice and are based upon a sensitivity and love for certain individual and general values prior to all axiological judgment. Some values appeal more to us than others, and if we listen to their calling and follow it, we enter into a unique personal relation with these values that is not subject to universal ethical judgment. At this level, it is not "wrong" to prefer values that are objectively lower ranked. This peculiar personal relation to values thus characterizes vocational life as a dimension of human life that is to some extent independent from the sphere of ethics.

Nevertheless, ethics is concerned with the way in which you live your vocational life and with its meaning for your life in general. Like Aristotle, Husserl thinks that the realm of ethics encompasses all kinds of vocational life in a universal manner. Moral demands do not determine which profession one should choose and are not concerned with the particular values and norms belonging to the different professions. But they do tell us how we should behave toward our professions and about the role that one's profession should have in one's life. The sphere of ethics concerns the universal form of human life as ruled by norms and ideals. These norms and ideals correspond to a particular kind of values, namely, moral values. The normative character of these values is fundamentally different from other kinds of values both as to the formal character of their prescription (which is unconditional or categorical) and as to the quali-

---

14 For a lucid discussion of Husserl's understanding of the relation between professional, vocational, and ethical life as three different forms of self-regulatory life, cf.: Heinämaa, 2014, pp. 198–203. Heinämaa convincingly points out that the thesis of a relation between ethics and a form of life is at the very heart of Husserl's approach to ethics. As the title of her essay suggests, Heinämaa focuses on Husserl's five *Kaizo* articles on renewal, three of which were published by Husserl himself in Japan in 1923 (see: Husserl, 1989). Here I will not make a distinction, as Heinämaa does, between professional and vocational life because I am following Husserl's reflections in *Grenzprobleme* and do not see such a distinction there. Husserl 2014, 388 f., 496, and 518.

15 Leisure activities are the counterpart of professional life, and are another field of our relation with values. They are subordinated to professional life but are a necessary part of a meaningful life and are connected with our "general profession" to exist as a human being. Husserl 2014, 388 f.

tative kind—the peculiar seriousness—of the ethical demand. To fail to obey ethical norms is of an incomparably different meaning and gravity than a failure to follow, e.g., the norms of aesthetics.

According to Husserl, ethical life is endowed with the teleological structure of a development toward the ideal of practical rationality. It is a life that aims both at contributing to a better world and at becoming a better person, shaping one's own personality in an ethical manner. In this respect, *critical reflection* and *deliberation* are of the utmost importance, and Husserl therefore insists upon their decisive role for ethical personhood.[16] Husserl thinks that the teleological tendency toward an ethical life is inherent in human nature and that only a life ruled by ethical rationality can be considered properly human.[17] In this light, there arises the question of how and to what extent it is possible *not* to live an ethical life and to ignore the demands of practical rationality. What does it mean to live a non-ethical life?

4 The possibility and structure of a non-ethical life

We pointed out that there are two different kinds of goods, those that can be enjoyed in this or that moment and those that possess a value that is not related to a certain temporary state, but to one's permanent being as a person and to one's life as a whole. This twofold order of possibilities and values corresponds to a twofold temporal structure of human existence. On the one hand, you live your life one present after another, living each present in an isolated manner. On the other hand, you always live your present as the present *of your life*, that is, as being part of your life in its entirety. Accordingly, there are goods that concern the present and yourself as the ego of the present; conversely, there are goods that concern your life as a whole and yourself as the ego of this life. These two kinds of goods can come into conflict.[18] If we want to understand this conflict, we must take into account the twofold manner in which a person lives her

16 For a detailed analysis of the attitude that characterizes an ethical life, see Heinämaa 2014, 203 ff. I am grateful to Thomas Nenon for having drawn my attention to the importance of reflection and deliberation in Husserl's ethics. On Husserl's account of the role of critical reflection for ethical personhood, see: Jacobs 2016, 323–335.

17 Cf., for instance, Husserl 2014, 321.

18 Cf.: Husserl 2014, 305, 395 f., and 491 f.

life: on the one hand, you live your life as a series of successively present episodes, each of which is lived in a certain discontinuity with respect to the rest of your life; on the other hand, you live each present as a present that continues your past and will be continued by your future.

The present is thus both in continuity and in discontinuity with one's life, and there is always a tension between these two ways in which we experience the present. This twofold nature of the present allows for two alternative attitudes. You can live primarily "in the present," understanding it in an isolated manner, or you can apprehend the present primarily as a part of your life. Correspondingly, you can identify yourself primarily with the ego of the present or with the ego of your life as a whole. Your ego is always at once the ego of a particular present—an ego that is different from one present to another—and the identical ego of your life. The identity and the difference of these two "egos" form an inseparable unity. You are always identical with and different from yourself, being at once the ego of the present moment and the ego of your life.[19]

This paradoxical temporal difference that is inseparable from the identity of the self gives rise to a certain division within the self that is also a division of its preferences. You normally prefer different things with regard to the present moment and with regard to your life. Therefore you always have to choose whether you want to act primarily as the ego of the present moment or as the ego of your life as a whole. Insofar as the ego of the present is also the ego of its entire life, it can relate to its personal identity in two alternative ways, namely, *one can be either true or untrue to oneself.*[20] You can live from the point of view of your permanent identity as the ego of your life and in this sense remain true to yourself, or you can "become untrue to yourself" by listening to your temporary preferences as the ego of your actual present. This analysis of the twofold temporality of the self and the tension between its continuity and discontinuity thus allows Husserl to account for the possibility of Aristotelian *akrasía* (incontinence) and for the Kantian opposition between duty and inclination.[21] The moral division within the self is made possible by the opposition between identity and self-difference that characterizes the temporal structure of human existence.

19 Husserl discusses the relation of the twofold nature of the living present as standing and streaming with the identity and self-difference of the ego in the C-manuscripts. This relation is the central theme of Held's *Lebendige Gegenwart*. Held 1966.

20 Husserl's expressions are "sich untreu werden" (become unfaithful to oneself) and "sich treu bleiben" (remain faithful to oneself); Husserl 2014, 227, 241, 453, 487 f., 491 f.

21 Husserl 2014, 265, 284, 292, 419 f.

If the tension between the present and one's entire life is habitually decided in favor of the former, this introduces a certain rupture within the ego that makes one lose control of one's life. Being unfaithful to oneself becomes a habit, and this motivates a loss of confidence in one's ability to carry out one's long-term decisions and projects. The result is a sort of *inactivity* (*Sich–treiben–lassen*, letting oneself drift)[22] in which we limit ourselves to following the demands of each moment. In this case, we give ourselves over to the present and its attractions, and this tendency becomes our habitual attitude toward life such that the primacy of the present characterizes the way in which we realize the temporal unity of the self.

Consequently, one could say that ethical life is not a universal and necessary form of human existence. Yet Husserl understands non-ethical life as a deficient mode or failure of living an ethical life that still is a certain form of ethical life. Non-ethical life is characterized by an intrinsic relation to ethical life and is no real alternative to it. In a certain sense, nobody can prefer a life that is not governed by ethical demands, because being true to oneself necessarily means living in accordance with the ideal of humanity, and this is the only way to live one's life as a temporal whole and as the life of an identical person. It is not only the case that an ethical life is the only truly satisfactory and happy life; but rather, there is simply no other way to live one's life as an identical person, because practical personal identity is inseparable from moral integrity. The ethical vocation is inscribed in human nature, and the ethical life is the only kind of life that is truly worth living. According to Husserl, a life that can be desired in its totality is thus necessarily an ethical life, whereas a life that is devoted to the goods of the present is not desirable as a whole, but only as the discontinuous sum of its successive whiles or moments.[23]

5 The External Conditions of Ethical Life

However, the possibility of living an ethical life depends not only on our decisions and actions, but also on other conditions. The first of these conditions concerns the structure of the world and its course. There must be a certain consistency and stability of our world of experience if we are to live a practical life

22 Husserl 2014, 265 and 455 f.
23 Husserl 2014, 487 f.

at all.[24] What is more, a fully satisfactory ethical life is only possible if the world allows for an infinite progression toward the ideal of a truly human world. The natural and social world must be sufficiently stable for the achievements of human progress to be conserved. Although this condition does not seem to be fulfilled for an infinite progression, Husserl claims that it is sufficient for an ethical life that the world is at least stable enough to allow for a temporary progression toward the ideal of humanity.[25]

The second and more specific external condition concerns the concrete course of the natural and social world in its relation to our ethical endeavors. It is possible that *fate* could annihilate the fruits of all individual and communal ethical effort. If we experience that all our effort is in vain or if a war or an accident destroys everything that we have been working for, the disappointment can reach such a degree that it must have a discouraging impact upon us and can even motivate complete resignation.[26] Nevertheless, Husserl thinks that it is possible to overcome this kind of discouragement in all cases. There is a meaning in ethical action that does not depend upon the success that it can expect.[27] Of course, and paradoxically, it is intrinsic to all action that we must be aiming at its success, but this success itself is neither the source nor the condition of the moral value of an action. On the contrary, this value is even higher if we have to face the blows of fate or the adversity of our circumstances. Another motive for consolation, closely connected to this, can be faith in the teleological structure of the world, in virtue of which everything ultimately serves a higher end and has a meaning, even if this meaning remains concealed to us. According to Husserl, this teleological structure is, so to speak, the worldly correlate of our faith in the value of ethical action, no matter how slight its visible effect might be. In a word, a certain amount of *good luck* is only a sufficient but not an absolutely necessary condition for the motivation of ethical action. Ideally speaking, there should be an unconditional faith in the meaning of ethical action, independently of its success.

Husserl believes that the teleological structure of the world is indeed a third condition for the meaningfulness of moral action and of an ethical life. Again,

24 Husserl 2014, 333.

25 Husserl 2014, 332. It may be interesting that in his lectures on ethics, Brentano repeatedly refers to Helmholtz's statement that only faith in the infinite progress of humanity could be a satisfactory motivation for an ethical life without faith in the soul's immortality. In a similar manner, Brentano himself claims that faith in the meaningfulness of our ethical effort ultimately depends upon faith in the existence of God. Cf.: Brentano 1952, 159 and 229 ff.

26 Husserl 2014, 304, 406 f.

27 Husserl 2014, 309 f.

this teleological structure is not an object of knowledge, but an object of faith. Nevertheless, knowledge and scientific convictions can invalidate this faith, and it is therefore a presupposition of an ethical life that the very meaning and possibility of such a life is not incompatible with our worldview. This kind of incompatibility can concern both our understanding of the world and our understanding of the human person. Thus, for example, the modern naturalistic worldview questions the meaningfulness of the world and thereby ultimately also questions the meaningfulness of human life and of ethical action. In its most radical form, it even directly questions the reality of voluntary action, declaring it an illusion and claiming to "reduce" it to mechanical processes.[28]

6 The Mission of Husserlian Transcendental Phenomenology

The correspondence between the teleology of human life and the structure of the world can only be assumed if the presuppositions of naturalistic objectivism can be overcome. The causal structure of the world can only correspond to the teleological structure of action if there is some kind of pre-established harmony.[29] Husserl believes that transcendental phenomenology is able to restore faith in this harmony, since it shows that the being-there both of the world of everyday life and of the natural world is the result of the teleological and meaningful process of constitution and is strictly correlated with the intersubjective and historical process of human life understood in a transcendental manner. What is more, the discovery of the dimension of transcendental constitution leads to the insight that it is mistaken to conceive of subjectivity as a product of the world, as something whose existence can be explained in terms of a natural causal process. Thus practical subjectivity regains its dignity, and the world regains its intrinsic relation to human practice. Of course, these brief indications can only become properly intelligible through a detailed study of Husserlian phenomenology.

28 Husserl 2014, 230 ff., 306.
29 Husserl 2014, 477 ff.

7 References

Brentano, F., 1952, *Grundlegung und Aufbau der Ethik.* Bern, Francke.

Brentano, F., 1955 [1889], *Vom Ursprung sittlicher Erkenntnis.* Hamburg, Meiner.

Heinämaa, S., 2014, Husserl's Ethics of Renewal: A Personalistic Approach, in: Tuominen, M., Heinämaa, S., and Mäkinen, V. (eds.), *New Perspectives on Aristotelianism and Its Critics.* Leiden, Brill, 196–212.

Held, K., 1966, Lebendige Gegenwart. Die Frage nach der Seinsweise des transzendentalen Ich bei Edmund Husserl, entwickelt am Leitfaden der Zeitproblematik. The Hague, Martinus Nijhoff.

Husserl, E., 1988, *Vorlesungen über Ethik und Wertlehre 1908–1914* (*Husserliana* XXVIII), ed. by U. Melle. Dordrecht, Kluwer.

Husserl, E., 1989, *Aufsätze und Vorträge* (*Husserliana* XXVII), ed. by Th. Nenon and H. R. Sepp. Dordrecht, Kluwer.

Husserl, E., 2004, *Einleitung in die Ethik* (*Husserliana* XXXVII), ed. by H. Peucker. Dordrecht, Kluwer.

Husserl, E., 2014, *Grenzprobleme der Phänomenologie* (*Husserliana* XLII), ed. by R. Sowa and Th. Vongehr. Dordrecht, Springer.

Jacobs, H., 2016, Socialization, Reflection, and Personhood, in: Rinofner-Kreidl, S. and Wiltsche, H. (eds.), *Analytic and Continental Philosophy: Methods and Perspectives; Proceedings of the 37th International Ludwig Wittgenstein Symposium.* Berlin, Walter de Gruyter, 323–335.

Palacios, J. M., 2003, Vorziehen und Wählen bei Scheler, in: Chr. Bermes, W. Henckmann, and H. Leonardy (eds.), *Vernunft und Gefühl.* Würzburg, Königshausen & Neumann, 135–146.

Palacios, J. M., 2008, Brentano en las inmediaciones del valor, in: Palacios, J. M., *Bondad moral e inteligencia ética.* Madrid, Encuentro, 31–44.

Scheler, M., 2000, Ordo amoris, in: *Schriften aus dem Nachlaß*, Band I (Ges. Werke, Bd. X), vierte Aufl. Bonn, Bouvier, 345–374.

Scheler, M., 2014, *Der Formalismus in der Ethik und die materiale Wertethik*, ed. by Christian Bermes and Annika Hand. Hamburg, Felix Meiner.

Abbreviations of Wittgenstein's Works

Bibliographical information regarding the edition used is given in the list of references attached to the relevant chapter. When a reference is to a numbered remark, this is indicated by a '§'; otherwise reference is to a page 'p.'. Abbreviations used to refer to works of other philosophers are given in the individual lists of references.

BB	The Blue and Brown Books
BT	The Big Typescript (TS 213)
CE	'Cause and Effect: Intuitive Awareness'
CV	Culture and Value
DB	Denkbewegungen: Tagebücher 1930–1932, 1936–1937
EPB	Eine Philosophische Betrachtung (Revision of parts of The Brown Book)
GT	Geheime Tagebücher
LC	Lectures and Conversations on Aesthetics, Psychology and Religious Belief
LE	'(Wittgenstein's) Lecture on Ethics'
LWPP I	Last Writings on the Philosophy of Psychology, Vol. 1
LWPP II	Last Writings on the Philosophy of Psychology, Vol. 2
NB	Notebooks 1914–1916
NFL	'Notes for Lectures on "Private Experience" and "Sense Data"'
OC	On Certainty
P	'Philosophy' (in BT/TS 213)
PG	Philosophical Grammar
PI	*Philosophical Investigations* (Part two is referred to as PI II, pg. nr.)
PO	Philosophical Occasions
PPO	Private and Public Occasions
PR	Philosophical Remarks
PTLP	*Prototractatus*: An Early Version of Tractatus Logico-Philosophicus
RC	Remarks on Colour
RFGB	'Remarks on Frazer's Golden Bough'
RFM	Remarks on the Foundations of Mathematics
RPP I	Remarks on the Philosophy of Psychology, Vol. 1
RPP II	Remarks on the Philosophy of Psychology, Vol. 2
TLP	Tractatus Logico-Philosophicus
WLPP	Wittgenstein's Lectures on the Philosophy of Psychology 1946–1947
Z	Zettel

https://doi.org/10.1515/9783110560282-179

References to Wittgenstein's manuscripts and typescripts are by MS and TS number following the G. H. von Wright catalogue in *Wittgenstein*, Oxford: Basil Blackwell, 1982; reprinted with an addendum in PO.

CL *Ludwig Wittgenstein: Cambridge Letters*, ed. Brian McGuinness and G. H. von Wright.
EL *Letters from Ludwig Wittgenstein–With a Memoir*, by Paul Engelmann, trans. L. Furtmüller, ed. Brian McGuinness.
FB *Familienbriefe*, ed. M. Ascher, B. McGuinness, and O. Pfersmann.
GB *Gesamtbriefwechsel*, ed. B. McGuinness, M. Seekircher, and A. Unterkircher.
LO *Letters to Ogden*, ed. Charles K. Ogden and G. H. von Wright.
LRKM Letters to Russell, Keynes and Moore, ed. G. H. von Wright.
WC Wittgenstein in Cambridge: Letters and Documents 1911–1951, ed. Brian McGuinness.
WE Wittgenstein–Engelmann: Briefe, Begegnungen, Erinnerungen

Notes taken by others

AWL Wittgenstein's Lectures, Cambridge 1932–3, ed. Alice Ambrose.
LC Lectures and Conversations on Aesthetics, Psychology and Religious Belief
LFM Wittgenstein's Lectures on the Foundations of Mathematics, Cambridge 1939, ed. Cora Diamond.
LSD 'The Language of Sense Data and Private Experience' (p. xiii)
MA Unpublished: Wittgenstein's lectures 1932–3 from the notes of G. E. Moore, Cambridge University Library G. E. Moore Archive, ref: ADD 8875, 10/7/7
MWL 'Wittgenstein's Lectures 1930–33', G. E. Moore's notes from Wittgenstein's lectures
NTW 'Notes on Talks with Wittgenstein'
VW The Voices of Wittgenstein: The Vienna Circle, Gordon Baker ed.
WL Wittgenstein's Lectures, Cambridge 1930–32, ed. Desmond Lee.
WLPP Wittgenstein's Lectures on the Philosophy of Psychology 1946–47, ed. Peter Geach.

WVC Wittgenstein and the Vienna Circle: Conversations recorded by Friedrich Waismann, ed. Brian McGuinness. (p. xiv).

Index

https://doi.org/10.1515/9783110560282-181

List of contributors

Prof. Dr. Kevin MULLIGAN
Université de Genève
Département de philosophie
Rue de Candolle 2
CH-1211 Genève 4 (SWITZERLAND)
e-mail: kevin.mulligan@unige.ch

Prof. Dr. Peter SIMONS
Trinity College Dublin
College Green
Dublin 2 (IRLAND)
e-mail: peter.simons@tcd.ie

Prof. Dr. Wolfgang KÜNNE
Universität Hamburg
Philosophisches Seminar
Von-Melle-Park 6
D-20146 Hamburg (GERMANY)
e-mail: wolfgang.kuenne@uni-hamburg.de

Prof. Dr. Christian BERMES
Lehrstuhl für Philosophie
Universität Koblenz-Landau
Institut für Philosophie
Bürgerstraße 23
D-76829 Landau (GERMANY)
e-mail: bermes@uni-landau.de

Prof. Dr. Michel LE DU
Département de philosophie
Faculté de Philosophie
7 Rue de L' Universite
F-67000 Strasbourg (FRANCE)
e-mail: ledu@unistra.fr

Prof. Dr. Bernhard OBSIEGER
Saint Louis University Madrid Campus
Avenida del Valle, 34
E-28003 Madrid (SPAIN)
e-mail: obsiegerb@slu.edu

https://doi.org/10.1515/9783110560282-185

Prof. Dr. Margit GAFFAL
Universidad Complutense, Madrid
Departamento de Filología Inglesa, I
Avenida de Filipinas, 3
E-28003 MADRID (SPAIN)
e-mail: mgaffal@hotmail.com

Prof. Dr. António MARQUES
Instituto de Filosofia da Nova
Faculdade de Ciências Sociais e Humanas
Universidade Nova de Lisboa
Av. de Berna, 26,4º Piso
1069-061 Lisboa (PORTUGAL)
e-mail: marquesantoni@gmail.com

Prof. Dr. Severin SCHROEDER
Department of Philosophy
Whiteknights Campus, HumSS G68
University of Reading
Reading RG6 6AA (UNITED KINGDOM)
e-mail: s.j.schroeder@reading.ac.uk

Prof. Dr. Amir HOROWITZ
Department of History, Philosophy and Judaic Studies
The Open University of Israel
1 University Road
P. O. Box 808
Raanana 43107 (ISRAEL)
e-mail: amirho@openu.ac.il

Prof. Dr. Jesús PADILLA GÁLVEZ
Universidad de Castilla-La Mancha
Facultad de Ciencias Jurídicas y Sociales
San Pedro Mártir, s/n
E-45071 Toledo (SPAIN)
e-mail: jesus.padilla@uclm.es